Edda Happ

Analyses of Evolutionary Algorithms

Edda Happ

Analyses of Evolutionary Algorithms

Theoretical Runtime Analyses for Evolutionary Algorithms and Probability Theoretical Tools for such Analyses

Südwestdeutscher Verlag für Hochschulschriften

Impressum / Imprint
Bibliografische Information der Deutschen Nationalbibliothek: Die Deutsche Nationalbibliothek verzeichnet diese Publikation in der Deutschen Nationalbibliografie; detaillierte bibliografische Daten sind im Internet über http://dnb.d-nb.de abrufbar.
Alle in diesem Buch genannten Marken und Produktnamen unterliegen warenzeichen-, marken- oder patentrechtlichem Schutz bzw. sind Warenzeichen oder eingetragene Warenzeichen der jeweiligen Inhaber. Die Wiedergabe von Marken, Produktnamen, Gebrauchsnamen, Handelsnamen, Warenbezeichnungen u.s.w. in diesem Werk berechtigt auch ohne besondere Kennzeichnung nicht zu der Annahme, dass solche Namen im Sinne der Warenzeichen- und Markenschutzgesetzgebung als frei zu betrachten wären und daher von jedermann benutzt werden dürften.

Bibliographic information published by the Deutsche Nationalbibliothek: The Deutsche Nationalbibliothek lists this publication in the Deutsche Nationalbibliografie; detailed bibliographic data are available in the Internet at http://dnb.d-nb.de.
Any brand names and product names mentioned in this book are subject to trademark, brand or patent protection and are trademarks or registered trademarks of their respective holders. The use of brand names, product names, common names, trade names, product descriptions etc. even without a particular marking in this work is in no way to be construed to mean that such names may be regarded as unrestricted in respect of trademark and brand protection legislation and could thus be used by anyone.

Verlag / Publisher:
Südwestdeutscher Verlag für Hochschulschriften
ist ein Imprint der / is a trademark of
OmniScriptum GmbH & Co. KG
Heinrich-Böcking-Str. 6-8, 66121 Saarbrücken, Deutschland / Germany
Email: info@svh-verlag.de

Herstellung: siehe letzte Seite /
Printed at: see last page
ISBN: 978-3-8381-1139-1

Zugl. / Approved by: Saarbrücken, Universität des Saarlandes, Diss., 2009

Copyright © 2009 OmniScriptum GmbH & Co. KG
Alle Rechte vorbehalten. / All rights reserved. Saarbrücken 2009

Abstract

Evolutionary algorithms (EAs) are a highly successful tool commonly used in practice to solve algorithmic problems. This remarkable practical value, however, is not backed up by a deep theoretical understanding. Such an understanding would facilitate the application of EAs to further problems. Runtime analyses of EAs are one way to expand the theoretical knowledge in this field.

This thesis presents runtime analyses for three prominent problems in combinatorial optimization. Additionally, it provides probability theoretical tools that will simplify future runtime analyses of EAs.

The first problem considered is the *Single Source Shortest Path* problem. The task is to find in a weighted graph for a given source vertex shortest paths to all other vertices. Developing a new analysis method we can give tight bounds on the runtime of a previously designed and analyzed EA for this problem.

The second problem is the *All-Pairs Shortest Path* problem. Given a weighted graph, one has to find a shortest path for every pair of vertices in the graph. For this problem we show that adding a crossover operator to a natural EA using only mutation provably decreases the runtime. This is the first time that the usefulness of a crossover operator was shown for a combinatorial problem.

The third problem considered is the *Sorting* problem. For this problem, we design a new representation based on trees. We show that the EA naturally arising from this representation has a better runtime than previously analyzed EAs.

Zusammenfassung

Evolutionäre Algorithmen (EAs) werden in der Praxis sehr erfolgreich eingesetzt. Bisher werden die theoretischen Grundlagen von EAs jedoch nicht zufriedenstellend verstanden. Laufzeitanalysen für einfache EAs sollen dieses Verständnis erweitern.

Diese Dissertation enthält Laufzeitanalysen für EAs für drei wohlbekannte kombinatorische Probleme. Zusätzlich werden wahrscheinlichkeitstheoretische Hilfsmittel zur Analyse von EAs eingeführt.

Zuerst behandeln wir das *Single Source Shortest Path* Problem. Die Aufgabe besteht darin, in einem gewichteten Graphen einen kürzesten Weg von einem Startknoten zu jedem anderen Knoten zu finden. Durch die Entwicklung einer neuen Analysemethode konnten wir scharfe Schranken für die Laufzeit eines bereits zuvor präsentierten und analysierten EAs angeben.

Als nächstes betrachten wir das *All-Pairs Shortest Path* Problem. Hierbei will man für jedes Paar von Knoten in einem gewichteten Graphen einen kürzesten Weg berechnen. Für dieses Problem zeigen wir, dass das Hinzufügen eines Crossover Operators die Laufzeit gegenüber einem natürlichen EA, der nur Mutationen nutzt, verbessert. Dies ist das erste Mal, dass für ein kombinatorisches Problem bewiesen wurde, dass ein Crossover Operator die Laufzeit reduziert.

Für das *Sortier*problem entwickeln wir eine neue, auf Bäumen beruhende Repräsentation und zeigen, dass der natürlich daraus entstehende EA eine bessere Laufzeit hat als vorherige EAs.

Acknowledgements

Many people have supported me during my work on this thesis.

First of all, I would like to thank my advisor Benjamin Doerr for his constant support and guidance at all times. He greatly inspired my interest in evolutionary algorithms. I am also very thankful to Kurt Mehlhorn for the opportunity to work in the inspiring research environment of his excellent research group at the Max-Planck-Institut für Informatik.

I would also like to thank all my present and former colleagues in this group who make working here so enjoyable. Of these, I want to mention especially my proofreaders Daniel Johannsen and Rolf Harren.

A great deal of the work reported in this thesis was a joint effort. For their collaboration and for their permission to include our results in my dissertation, I am very grateful to my coauthors Benjamin Doerr and Christian Klein.

Contents

1 Introduction — 1
 1.1 The Paradigm of Evolutionary Algorithms 1
 1.2 Randomized Search Heuristics 2
 1.3 Theoretical Analysis 3
 1.4 Overview 4

2 Evolutionary Algorithms — 7
 2.1 Components of an Evolutionary Algorithm 7
 2.2 Individuals and Population 7
 2.3 Fitness Function and Selection Operator 8
 2.4 Variation Operators 9
 2.5 Structure of an Evolutionary Algorithm 10
 2.6 Optimization Time 11

3 Probability Theory — 13
 3.1 Chernoff Bounds 13
 3.2 Application of Chernoff Bounds for Correlated Variables . . . 15
 3.3 Derivation of Expected Values from High Probabilities 17

4 The Single Source Shortest Path Problem — 19
 4.1 Introduction 19
  4.1.1 Related Work 19
  4.1.2 Our Results 20
 4.2 An Evolutionary Algorithm for the SSSP Problem 20
  4.2.1 Individuals 21
  4.2.2 Fitness Function 22
  4.2.3 Mutation Operator 22
  4.2.4 The $(1+1)$-EA$_{sssp}$ 22
 4.3 Upper Bound on the Optimization Time 23
 4.4 Lower Bound on the Optimization Time 27

		4.4.1 A Worst Case Graph Class	27
		4.4.2 A Lower Bound	28
	4.5	Summary	35

5 The All-Pairs Shortest Path Problem — 37
- 5.1 Introduction — 37
  - 5.1.1 Related Work — 37
  - 5.1.2 Our Results — 39
- 5.2 A Genetic Algorithm for the APSP Problem — 40
  - 5.2.1 Individuals and Population — 41
  - 5.2.2 Fitness Function and Selection Operator — 41
  - 5.2.3 Mutation and Crossover Operators — 42
  - 5.2.4 The $(\mu+1)$-EA$_{apsp}$ and $(\mu+1)$-GA$_{apsp}$ — 43
- 5.3 Analysis of the $(\mu+1)$-EA$_{apsp}$ — 45
  - 5.3.1 Upper Bound on the Optimization Time — 45
  - 5.3.2 Lower Bound on the Optimization Time — 48
- 5.4 Upper Bound on the Optimization Time of the $(\mu+1)$-GA$_{apsp}$ — 52
- 5.5 Experimental Results — 58
- 5.6 Summary — 63

6 Sorting — 65
- 6.1 Introduction — 65
  - 6.1.1 Related Work — 66
  - 6.1.2 Our Results — 67
- 6.2 An Evolutionary Algorithm for Sorting — 69
  - 6.2.1 Individuals — 70
  - 6.2.2 Fitness Function — 70
  - 6.2.3 Mutation Operator — 71
  - 6.2.4 The $(1+1)$-EA$_{sort}$ — 72
- 6.3 Analysis of the Optimization Time — 73
- 6.4 Implementation Details and Analysis of the Actual Runtime — 77
- 6.5 Experimental Results — 80
- 6.6 Summary — 86

7 Summary — 87

A Further Contributions — 89

Bibliography — 90

List of Figures

2.1	Structure of an EA	11
4.1	SSSP: Graph and vector representation of an individual	21
4.2	SSSP: The $(1+1)$-EA$_{sssp}$	23
4.3	SSSP: Graph class for the lower bound – shortest path tree	27
4.4	SSSP: Example graph for the lower bound	27
5.1	APSP: The three crossover operators	42
5.2	APSP: The $(\mu+1)$-GA$_{apsp}$	44
5.3	APSP: Example graph K_n, w for the lower bound	48
5.4	APSP: Example graph K_n, w' for the $\otimes_2$-operator	57
5.5	APSP: Example graph K_n'', w'' for the $\otimes_3$-operator	57
5.6	APSP: Sets of shortest paths in K_n'', w''	58
5.7	APSP: Optimization times for K_n, w	59
5.8	APSP: Optimization times for K_n, w'	60
5.9	APSP: Optimization times for K_n'', w''	61
5.10	APSP: Log-log plots for K_n, w	61
5.11	APSP: Log-log plots for K_n, w'	62
5.12	APSP: Log-log plots for K_n'', w''	62
5.13	APSP: Exponents of the optimization times	63
6.1	Sorting: The initial solution	67
6.2	Sorting: A permutation	68
6.3	Sorting: An elementary mutation	68
6.4	Sorting: The $(1+1)$-EA$_{sort}$	73
6.5	Sorting: The datastructures used by the $(1+1)$-EA$_{sort}$	78
6.6	Sorting: The datastructure siblings	79
6.7	Sorting: Optimization times of the $(1+1)$-EA$_{sort}$	82
6.8	Sorting: Optimization times of the $(1+1)$-EA$_p$ (i)	83
6.9	Sorting: Optimization times of the $(1+1)$-EA$_p$ (ii)	84
6.10	Sorting: Optimization times – $(1+1)$-EA$_{sort}$ vs. $(1+1)$-EA$_p$	85

Chapter 1
Introduction

Evolutionary algorithms are a highly effective and widely used tool for solving a broad range of optimization problems. The utilization of evolutionary algorithms is very easy and they quickly provide results of a high quality. But what exactly are evolutionary algorithms and why are they so efficient?

In this thesis, we introduce evolutionary algorithms and present theoretical results contributing to the answer of the second question.

1.1 The Paradigm of Evolutionary Algorithms

Evolutionary algorithms use the principles of biological evolution in an algorithmic framework to solve problems from computer science.

Charles Darwin is generally seen as the father of the evolutionary thought. His new evolutionary theory was based on natural variation and selection. This, in combination with Gregor Mendel's laws of inheritance, gave rise to the design of several distinct problem solving techniques in computer science, comprised under the general term of evolutionary computation or evolutionary algorithms.

Biological evolution is a process spread over many generations that gradually changes the inheritance traits of a population of individuals. These traits are encoded in the genes and vary within the population. Mutation changes the genes over time, and sexual reproduction combines genes of different individuals. By natural selection, traits helpful for survival and reproduction become more common in the population, whereas harmful ones extinguish.

This occurs because individuals with advantageous traits have better chances to pass these traits on to the next generation (survival of the fittest).

Evolutionary algorithms use computational models of these evolutionary processes for randomized computational optimization. Typically, they keep a set (population) of solution candidates (individuals), which they try to gradually improve. Improvements may be generated by applying different variation operators, such as mutation and recombination (mostly called crossover), to certain individuals. The quality of a solution is measured by a so-called fitness function. Based on the fitness value of the individuals, a selection procedure removes some individuals from the population. The cycle of variation and selection is repeated until a solution of sufficient fitness is found. See, e.g., [For93] for a short introduction to genetic algorithms. In Section 2 we will define evolutionary algorithms more thoroughly.

1.2 Randomized Search Heuristics

Evolutionary algorithms are successfully used for a wide range of optimization problems. They belong to the class of randomized search heuristics [Weg03]. Among others, *Randomized Local Search* (RLS), the *Metropolis Algorithm* [MRR+53], and *Simulated Annealing* [KGJV83] also belong to this class of algorithms. All of them try to find good solutions to an optimization problem by repeatedly changing some candidate solution(s) randomly.

Whereas early hopes that these ideas might make notoriously hard problems become tractable were not fulfilled [WM97], randomized search heuristics nowadays are frequently used as a generic way to obtain algorithms. Naturally, such generic approaches cannot compete with a custom-tailored algorithm. Nevertheless, they are a well established tool that is frequently used in practice, because they are easy and cheap to implement, need fewer analysis of the problem to be solved, and can be reused easily for related problems. An expert in such methods can comfortably solve algorithmic problems by plugging together suitable generic components without fully analyzing the problem itself. The components only have to be adapted to the particular problem under consideration. This adaptation can be guided by an experimental evaluation of the actual behavior of the algorithms or by previously obtained experience.

1.3 Theoretical Analysis

Though evolutionary approaches have proven to be extremely successful in practice (see, e. g., the Proceedings of the annual ACM Genetic and Evolutionary Computation Conferences (GECCO)), the theoretical understanding of such methods is still in its infancy.

Nevertheless, the recent years produced some remarkable theoretical results, mostly on convergence phenomena and runtime analyses. Since we will present runtime analyses, we point the reader interested in some convergence results to [RW91, RRS95, RW99]. The aim of runtime analyses is to obtain a theoretically founded understanding of the basic principles of evolutionary computation. The hope is that such an understanding will facilitate the future design of evolutionary algorithms.

The first theoretical runtime analyses were conducted on simple evolutionary algorithms on artificial problems like maximizing simple pseudoboolean functions $f \colon \{0,1\}^n \to \mathbb{R}$, e. g., the number of ones (ONEMAX$(x) := \sum_{i=1}^{n} x_i$), the number of leading ones (LO$(x) := \max\{i \in \mathbb{N} \mid \forall j \leq i : x_j = 1\}$), or monotone linear functions and polynomials [Weg01, DJW02, WW05]. Already through the work on these simple problems quite some insight was gained on the functionality and analysis of evolutionary algorithms. Surprisingly, even for such extremely simple evolutionary algorithms on simple problems a tight analysis of the runtime behavior can be very complicated. A classical example for the difficulties one faces when analyzing evolutionary algorithms is the (tight) $O(n \log(n))$ bound for the optimization time of a simple $(1+1)$ evolutionary algorithm maximizing a monotone linear function on $\{0,1\}^n$. Here, classical methods from the analysis of randomized algorithms lead to a highly technical proof [DJW02]. Subsequent efforts put into this problem resulted in the so-called drift analysis becoming a major tool in the runtime analysis of evolutionary algorithms [HY04].

More recently, evolutionary algorithms for classical problems from computer science became the focus of such runtime analyses. The first work in this direction was conducted by Scharnow, Tinnefeld, and Wegener who analyzed simple evolutionary algorithms for sorting and shortest path problems [STW04]. Results on evolutionary algorithms for combinatorial problems like the Eulerian cycle problem [Neu04, DHN06, DKS07, DJ07], minimum spanning trees [NW04, NW05], maximal matchings [GW03], and partition [Wit05] followed. The design of evolutionary algorithms for such combinatorial problems becomes more interesting, since one typically has to choose between several natural representations of the individuals, fitness functions, and variation operators. Naturally, because of the richer structure of such problems, the analysis becomes also more challenging.

To avoid misunderstandings, let us stress that the focus of this line of research is not to find superior algorithms for the particular underlying optimization problem. Since these are classical and important problems, they have been investigated thoroughly and hence very good custom-tailored algorithms already exist. Rather, the focus of this work is to analyze how such problems can be tackled with generic approaches, to understand how their components like particular representations or variation operators work, and, finally, to develop methods to analyze evolutionary algorithms.

1.4 Overview

The following gives a brief overview of the rest of this work summarizing our main contributions and referencing the publications this thesis is based on.

Chapter 2 explains evolutionary algorithms in more detail, discussing their components and structure. Chapter 3 presents the necessary background on probability theory. This includes classical Chernoff bounds as well as some tools we developed for our analyses.

In Chapter 4, we reconsider the evolutionary algorithm for the Single Source Shortest Path problem introduced in [STW04]. We apply a new technique for the analysis that overcomes the coupon collector behavior usually used. Using this technique, we improve the previous results by giving a tight bound on the optimization time which holds with high probability. Based on the paper "A Tight Analysis of the $(1 + 1)$-EA for the Single Source Shortest Path Problem" by B. Doerr, E. H., and C. Klein [DHK07].

In Chapter 5, we consider the All-Pairs Shortest Path problem and design a natural evolutionary algorithm for it. We rigorously analyze the optimization time of this algorithm with and without a crossover operator, which reveals that adding a crossover operator can in fact improve the optimization time. This is the first time that the usefulness of a crossover operator was shown for a combinatorial problem. Based on the paper "Crossover Can Provably be Useful in Evolutionary Computation" by B. Doerr, E. H., and C. Klein [DHK08].

In Chapter 6, we introduce a simple framework for dealing with search spaces consisting of permutations. To demonstrate its usefulness, we build upon it a simple evolutionary algorithm for the Sorting problem. We show that this algorithm has a better optimization time than the currently best

1.4 Overview

evolutionary algorithm for the Sorting problem introduced in [STW04]. Additionally, our approach has the particular advantage that it does distinguish between wrong and unexplored information. This allows to retrieve partial, correct information even before the optimal solution has been found. Based on the paper "Directed Trees: A Powerful Representation for Sorting and Ordering Problems" by B. Doerr and E. H. [DH08].

The Appendix A lists further contributions that are not contained in this thesis.

Chapter 2
Evolutionary Algorithms

The paradigm of evolutionary computation is to use principles inspired by the biological evolution (cf. Section 1.1) to find good solutions for optimization problems. In this chapter, we explain the different components and the general structure of an evolutionary algorithm (EA).

2.1 Components of an Evolutionary Algorithm

The aim of an evolutionary algorithm is to find an optimal point in a search space. It does so by keeping an over time evolving *population* (a set) of *individuals* (or candidate solutions). These individuals represent points in the search space. *Variation operators*, such as *mutation* and *crossover*, are used to create new individuals from the existing population. As in nature, the size of the population is not supposed to grow infinitely, and thus some of the individuals are selected to persist, whilst others are deleted. There exist several *selection strategies* which are typically based on a *fitness function* which assigns fitness values to the individuals.

In the following, the different components of an evolutionary algorithm are explained in more detail.

2.2 Individuals and Population

An evolutionary algorithm tries to find an optimal solution in a search space $\mathcal{S}$. Hence, an individual that is a candidate solution to the considered optimization problem has to be a point in $\mathcal{S}$. In the beginning of runtime analyses for evolutionary algorithms, EAs optimizing pseudo-boolean functions were considered. Here, the search space is $\mathcal{S} := \{0,1\}^h$ and the individuals are bit-strings of a fixed length h. However, if more elaborate problems are considered (as in this work), it is usually more adequate to choose a different representation (and thus search space) for the problem.

The set of currently considered individuals is called the population. The population at time t is denoted by $\mathcal{I}^t$, and the n_t individuals in this population are denoted by $I_1^t, \ldots, I_{n_t}^t$. The index t is often omitted if it does not create ambiguity. A population at a certain time step is sometimes also called the *current population* or *generation*. When talking of populations at time t and $t+1$ one talks about the *parent* and the *offspring* population or generation. The size of the population is usually denoted by μ and the number of offspring individuals by λ. In many cases, the size of the population and the number of offspring individuals created per generation is constant. When analyzing evolutionary algorithms, often (as in this work) simple versions with $\mu = 1$ and / or $\lambda = 1$ are considered.

2.3 Fitness Function and Selection Operator

The aim of the selection operator is to prevent the population from growing too big as well as to get rid of individuals that are not considered to be useful solution candidates anymore. Typically, selection is guided by a fitness function $f \colon \mathcal{S} \to \mathbb{R}$ assigning each individual $I \in \mathcal{I}$ a fitness value. This fitness function is a heuristic measure that indicates how far an individual is from being optimal.

An evolutionary algorithm is called a (μ, λ)-EA if it has a population size of μ, creates λ offspring individuals, and of these offspring individuals selects the μ individuals having the best fitness. If the fittest μ individuals are instead selected from a union of the parent population and the offspring, the algorithm is called a $(\mu + \lambda)$-EA. The latter is more commonly used and the one we consider in this work.

Both of the above described selection strategies choose the fittest individuals and are thus called elitist selection strategies (or truncation). There also exist other selection strategies [GD91, BT96], e.g., fitness-proportional (also called roulette-wheel) selection and tournament selection. Fitness-proportional selection chooses the individuals for the next generation with a probability proportional to their fitness values. Tournament selection chooses τ (called the tournament size) individuals at random of which it selects the fittest. This process is repeated as often as individuals must be chosen.

2.4 Variation Operators

Elitist selection can lead to a faster fitness increase in the population, whereas fitness-proportional and tournament selection have the advantage of a higher degree of diversity in the population. The advantage of diversity in a population is that the individuals hopefully cover a bigger part of the search space and thus not all get stuck in some local optimum. A mechanism that ensures variety in a population is called a diversity mechanism.

We do not need diversity mechanisms in Chapters 4 and 6, since there we analyze $(1+1)$-EAs which always have a population consisting of a single individual. In Chapter 5, diversity is assured by different measures using elitist selection, so that we will not use any other selection strategies.

2.4 Variation Operators

Mutation and crossover are used to generate new individuals. Mutation does so by applying some random changes to one individual whilst crossover randomly combines two individuals. In this work, we will call an algorithm that uses only mutation an evolutionary algorithm (EA), and one that uses mutation and crossover a genetic algorithm (GA). In practice, mutation is often applied to the individual resulting from a crossover step.

Mutation Operator

The classical mutation operator for bit-strings of length h selects a random individual $I := (x_1, \ldots, x_h) \in \mathcal{I} \subseteq \{0,1\}^h$ from the population and creates a new individual I_{new} by flipping each bit of I with probability $\frac{1}{h}$. By doing this, in expectation one bit is flipped. However, it is possible, though improbable, for a single mutation step to flip many bits, including up to all h bits. Thus, any bit-string might be generated by a single mutation step, but usually a bit-string in a closer neighborhood is created. This is in contrast to RLS (Randomized Local Search) where only a close neighborhood can be reached in one step [WW03].

If the individuals are not represented by bit-strings, a different mutation operator fitting the representation has to be defined. Analogously to the classical one flipping a number of bits, the mutation operator is supposed to apply a number of slight random changes to an individual. Let us call such a slight change an *elementary mutation*. The number of bits flipped by the classical mutation operator can be approximated by using a Poisson distribution $\mathrm{Pois}(\zeta = 1)$ with parameter $\zeta = 1$, as proposed in [STW04].

The Poisson distribution with $\zeta = 1$ is the limit (for h growing large) of the Binomial distribution for h trials with probability $\frac{1}{h}$ each. For general $\zeta > 0$ and $k \in \mathbb{N}$, the Poisson distribution with parameter ζ is given by the probability mass function

$$f(k;\zeta) = \frac{\zeta^k e^{-\zeta}}{k!},$$

which gives the probability that a $\mathrm{Pois}(\zeta)$-distributed random variable is equal to k. The number of elementary mutations applied in one step is then given by $S+1$ (to assure that at least one elementary mutation is performed), where S is distributed according to $\mathrm{Pois}(\zeta = 1)$.

Crossover Operator

The classical crossover operators for bit-strings are *uniform* and *one-point* crossover. Both randomly select two individuals I_i and I_j from the population $\mathcal{I}$ and randomly create a new individual I_{new}. The uniform crossover does so by setting for $k \in [1..h]$ the k-th bit of I_{new} with probability $\frac{1}{2}$ to the k-th bit of either parent. The one-point crossover chooses a position $k \in [0..h]$ uniformly at random and sets the first k bits of I_{new} to the first k bits of I_i and the last $h - k$ bits to the corresponding bits of I_j.

If a different representation than bit-strings is used, a crossover operator fitting the representation has to be defined that combines two individuals in a random manner.

2.5 Structure of an Evolutionary Algorithm

Typically, an evolutionary algorithm (EA) has the following structure (cf. Figure 2.5). First the initial population is created, often this initialization is done randomly. Until a solution of sufficient fitness is found, the evolutionary algorithm repeats the following steps. A number (usually denoted by λ) of variation steps are done to generate (λ) offspring individuals. The variation step consists of a mutation and / or a crossover step. If crossover is applied, we will call the algorithm a genetic algorithm (GA). A selection step then selects the (μ) individuals that form the next generation.

2.6 Optimization Time

```
GENERAL EA
1    Initialization                              (create initial population $\mathcal{I}$)
2    repeat
3         repeat $\lambda$ times
4              Variation                        (create a new individual)
5              Selection                        (choose $\mu$ individuals)
6    until $\mathcal{I}$ contains solution of sufficient fitness
```

Figure 2.1: The general structure of an evolutionary algorithm.

2.6 Optimization Time

When analyzing the complexity of an algorithm, one is typically interested in the asymptotic runtime of the algorithm, that is, constant factors and lower order terms are ignored. Since evolutionary algorithms are randomized algorithms, we can only consider the runtime as random variable and analyze the expected value of this random variable.

Moreover, the preferred performance measure in the EA community is not the runtime, but the *optimization time*. This is defined to be the number of fitness function evaluations the algorithm performs until it finds a solution of sufficient fitness, in this thesis an optimal solution. The time needed for the creation of new individuals by mutation or crossover or to evaluate the fitness function is usually disregarded. Again, only the expected asymptotic behavior is of interest. If the number of offspring individuals created per generation is constant, the asymptotic optimization time is equal to the asymptotic number of generations needed to find the desired solution.

However, not always is the expected optimization time the only term of interest when analyzing evolutionary algorithms. Sometimes one wants to state additionally to or instead of the expected optimization time an upper or lower bound on the optimization time that holds up to a certain failure probability. A bound on this failure probability is typically inverse polynomial or inverse exponential. We say an event holds *with high probability* if it holds with probability at least $1 - O(n^{-c})$ for an arbitrary but fixed constant c. Equivalently, we say an event holds *with overwhelming probability* if it holds with probability at least $1 - 2^{-\Omega(n^{\varepsilon})}$ for some constant $\varepsilon > 0$.

Chapter 3
Probability Theory

Since evolutionary algorithms are random algorithms, for the analysis some background in probability theory is needed. In this chapter we introduce some classical Chernoff bounds together with several theorems and lemmas that we developed and used while working on the contributions presented in this thesis.

3.1 Chernoff Bounds

To be able to state that an optimization time holds with high or overwhelming probability (cf. Section 2.6), throughout this work we will often use the following classical bounds on large deviations [AS00, MR95].

Theorem 3.1 (Chernoff Bounds). *Let $X_1, \ldots, X_t$ be mutually independent random variables with $X_i \in \{0, 1\}$ for all $i \in [1..t]$. Let $X := \sum_{i=1}^{t} X_i$. Then*

a) for all $\alpha < 1$, $\Pr[X < \alpha \mathbb{E}[X]] \leq \exp(-\frac{1}{2}(1-\alpha)^2 \mathbb{E}[X])$,

b) for all $\beta > 1$, $\Pr[X \geq \beta \mathbb{E}[X]] < (e^{\beta-1} \beta^{-\beta})^{\mathbb{E}[X]}$,

c) for all $\gamma > 0$, $\Pr[X \geq (1+\gamma)\mathbb{E}[X]] \leq \exp\left(-\frac{\min\{\gamma, \gamma^2\} \mathbb{E}[X]}{3}\right)$.

If the variables are geometrically distributed, the following Chernoff-like inequality can be helpful. To the best of our knowledge, such bounds have not been published so far in a mathematics or computer science journal.

Theorem 3.2. Let $Y_1, \ldots, Y_t$ be mutually independent random variables, $Y := \sum_{i=1}^{t} Y_i$, and $0 < p < 1$ be a constant. If the Y_i are geometrically distributed random variables with $\Pr[Y_i = j] = (1-p)^{j-1}p$ for all $j \in \mathbb{N}$, then

for all $\delta > 0$, $\Pr[Y > (1+\delta)\mathbb{E}[Y]] \leq \exp\left(-\frac{\delta^2}{2}\frac{(t-1)}{(1+\delta)}\right)$.

Proof. Let $X_1, X_2, \ldots$ be an infinite sequence of independent, identically distributed biased coin tosses (binary random variables) such that X_i is one with probability $\Pr[X_i = 1] = p$ and zero with probability $\Pr[X_i = 0] = 1 - p$. Note that the random variable "smallest j such that $X_j = 1$" has the same distribution as each Y_i. In consequence, Y has the same distribution as "smallest j such that exactly t of the variables $X_1, \ldots, X_j$ are one". In particular, $\Pr[Y > j] = \Pr[\sum_{i=1}^{j-1} X_i < t - 1]$ for all $j \in \mathbb{N}$. This manipulation reduces our problem to the analysis of independent Bernoulli trials and will enable us to use the classical Chernoff bounds.

The expected value of each Y_i is $\mathbb{E}[Y_i] = \frac{1}{p}$, thus $\mathbb{E}[Y] = \frac{t}{p}$. Let $X := \sum_{i=1}^{\lceil(1+\delta)\mathbb{E}[Y]\rceil-1} X_i$. By the above,

$$\Pr[Y > (1+\delta)\mathbb{E}[Y]] = \Pr[X < t - 1].$$

The expected value of X is bounded by

$$\mathbb{E}[X] = \lceil(1+\delta)\mathbb{E}[Y] - 1\rceil p \geq (1+\delta)t - p > (1+\delta)(t-1).$$

Now let $\alpha := \frac{t-1}{\mathbb{E}[X]}$. Then $\alpha < 1$ and $\Pr[X < t - 1] = \Pr[X < \alpha\mathbb{E}[X]]$. Hence we can apply the first inequality in Theorem 3.1 to get

$$\begin{aligned}
\Pr[Y > (1+\delta)\mathbb{E}[Y]] &= \Pr[X < \alpha\mathbb{E}[X]] \\
&\leq \exp\left(-\frac{1}{2}\mathbb{E}[X](1 - \frac{t-1}{\mathbb{E}[X]})^2\right) \\
&\leq \exp\left(-\frac{1}{2}\mathbb{E}[X](1 - \frac{1}{1+\delta})^2\right) \\
&\leq \exp\left(-\frac{1}{2}(t-1)(1+\delta)(\frac{\delta}{1+\delta})^2\right) \\
&= \exp\left(-\frac{\delta^2}{2}\frac{(t-1)}{(1+\delta)}\right).
\end{aligned}$$

□

3.2 Application of Chernoff Bounds for Correlated Variables

With Chernoff Bounds we can handle sums of independent random variables. In our proofs, however, we will also encounter sums of correlated variables, since in evolutionary algorithms events usually depend on the steps the algorithm has performed so far. To be able to deal with sums of correlated variables we will use the following lemma. With it, we can approximate the behavior of such sums by using sums of independent random variables. Hence we can, albeit indirectly, apply Chernoff Bounds to certain sums of dependent random variables.

Lemma 3.1. *Let $X_1, \ldots, X_t, X_1^*, \ldots, X_t^*$ be random variables that may take on natural numbers as values. For all $i \in [1..t]$, X_i is independent of $X_{i+1}, \ldots, X_t$, and the $X_1^*, \ldots, X_t^*$ are mutually independent. Then for all $k \geq 0$ the following holds.*

a) *If for all $i \in [1..t]$, all $m \in \mathbb{N}^+$, and all $x_1, \ldots x_{i-1} \in \mathbb{N}$*

$$\Pr[X_i = m \mid X_1 = x_1, \ldots, X_{i-1} = x_{i-1}] \geq \Pr[X_i^* = m],$$

then

$$\Pr[\sum_{i=1}^{t} X_i \geq k] \geq \Pr[\sum_{i=1}^{t} X_i^* \geq k].$$

b) *If for all $i \in [1..t]$, all $m \in \mathbb{N}^+$, and all $x_1, \ldots x_{i-1} \in \mathbb{N}$*

$$\Pr[X_i = m \mid X_1 = x_1, \ldots, X_{i-1} = x_{i-1}] \leq \Pr[X_i^* = m],$$

then

$$\Pr[\sum_{i=1}^{t} X_i \geq k] \leq \Pr[\sum_{i=1}^{t} X_i^* \geq k].$$

Proof. a) Denote by $P_j := \Pr[\sum_{i=1}^{j} X_i + \sum_{i=j+1}^{t} X_i^* \geq k]$ for $j \in [0..t]$ the probability that given the sequence of events $X_1, \ldots, X_j, X_{j+1}^*, \ldots, X_t^*$ the sum $\sum_{i=1}^{j} X_i + \sum_{i=j+1}^{t} X_i^*$ is at least k. Then we get for P_j for $j \in [1..t]$

$$P_j = \Pr\left[\sum_{i=1}^{j} X_i + \sum_{i=j+1}^{t} X_i^* \geq k\right]$$

$$= \Pr\left[\sum_{i=1}^{j-1} X_i + \sum_{i=j+1}^{t} X_i^* \geq k\right] \cdot 1 +$$

$$\sum_{m=1}^{k}\left(\Pr\left[\sum_{i=1}^{j-1} X_i + \sum_{i=j+1}^{t} X_i^* = k-m\right] \cdot \Pr[X_j \geq m]\right)$$

$$= \Pr\left[\sum_{i=1}^{j-1} X_i + \sum_{i=j+1}^{t} X_i^* \geq k\right] +$$

$$\sum_{m=1}^{k} \sum_{\substack{(x_1,\ldots,x_{j-1},x_{j+1},\ldots,x_t) \\ \in \mathcal{X}_{k-m}^{t-1}}} \left(\prod_{i=1}^{j-1} \Pr\left[X_i = x_i \mid X_1 = x_1, \ldots, X_{i-1} = x_{i-1}\right] \cdot \right.$$

$$\left. \prod_{i=j+1}^{t} \Pr\left[X_i^* = x_i\right] \cdot \Pr\left[X_j \geq m \mid X_1 = x_1, \ldots, X_{j-1} = x_{j-1}\right]\right)$$

where $\mathcal{X}_{k-m}^{t-1} = \{(x_1,\ldots,x_{t-1}) \in \mathbb{N}^{t-1} \mid \sum_{i=1}^{t} x_i = k-m\}$ is the set of event outcomes $(x_1,\ldots,x_{t-1}) \in \mathbb{N}^{t-1}$ that fulfil $\sum_{i=1}^{t-1} x_i = k-m$. Using the minimum of $\Pr[X_j \geq m \mid X_1 = x_1,\ldots,X_{j-1} = x_{j-1}]$ over all $(x_1,\ldots,x_{j-1}) \in \mathbb{N}^{j-1}$ that gives

$$P_j \geq \Pr\left[\sum_{i=1}^{j-1} X_i + \sum_{i=j+1}^{t} X_i^* \geq k\right] +$$

$$\sum_{m=1}^{k}\left(\Pr\left[\sum_{i=1}^{j-1} X_i + \sum_{i=j+1}^{t} X_i^* = k-m\right] \cdot \min_{(x_1,\ldots,x_{j-1}) \in \mathbb{N}^{j-1}} \Pr\left[X_j \geq m \mid X_1 = x_1,\ldots,X_{j-1} = x_{j-1}\right]\right).$$

Since for all $j \in [1..t]$, $m \in \mathbb{N}^+$, and $x_1,\ldots,x_{j-1} \in \mathbb{N}$ we have that $\Pr[X_j = m \mid X_1 = x_1,\ldots,X_{j-1} = x_{j-1}] \geq \Pr[X_j^* = m]$, we get that $\Pr[X_j \geq m \mid X_1 = x_1,\ldots,X_{j-1} = x_{j-1}] \geq \Pr[X_j^* \geq m]$, and thus

$$P_j \geq \Pr\left[\sum_{i=1}^{j-1} X_i + \sum_{i=j+1}^{t} X_i^* \geq k\right] +$$
$$\sum_{m=1}^{k} \left(\Pr\left[\sum_{i=1}^{j-1} X_i + \sum_{i=j+1}^{t} X_i^* = k - m\right] \cdot \Pr\left[X_j^* \geq m\right]\right)$$
$$= \Pr\left[\sum_{i=1}^{j-1} X_i + \sum_{i=j}^{t} X_i^* \geq k\right]$$
$$= P_{j-1}.$$

Thus, we have that $\Pr[\sum_{i=1}^{t} X_i \geq k] = P_t \geq P_{t-1} \geq \cdots \geq P_1 \geq P_0 = \Pr[\sum_{i=1}^{t} X_i^* \geq k]$.

b) Let $P_j := \Pr[\sum_{i=1}^{j} X_i + \sum_{i=j+1}^{t} X_i^* \geq k]$ for $j \in [0..t]$ as above. Using the maximum instead of the minimum over all $(j-1)$-tuples and the fact that $\Pr[X_j = m \mid X_1 = x_1, \ldots, X_{j-1} = x_{j-1}] \leq \Pr[X_j^* = m]$ and thus $\Pr[X_j \geq m \mid X_1 = x_1, \ldots, X_{j-1} = x_{j-1}] \leq \Pr[X_j^* \geq m]$ for all $j \in [1..t]$, $m \in \mathbb{N}^+$, and $x_1, \ldots, x_{j-1} \in \mathbb{N}$ in the calculations from a), we get that $P_j \leq P_{j-1}$ for $j \in [1..t]$. Hence it follows that $\Pr[\sum_{i=1}^{t} X_i \geq k] \leq P_{t-1} \leq \cdots \leq P_1 \leq \Pr[\sum_{i=1}^{t} X_i^* \geq k]$. □

Note that this lemma also holds if the considered variables X_i and X_i^* are binary random variables and thus $m = 1$.

3.3 Derivation of Expected Values from High Probabilities

Throughout this work we often use Chernoff inequalities to derive bounds on the optimization time that hold with high or overwhelming probability. The following lemmas allow to deduce thereof bounds on the expected optimization time.

Lemma 3.2. *Let $c_2 > 0$, $\eta' > 0$, and c_1 be constants, let $n \in \mathbb{N}^+$, and let $g(n) > 0$ be a function. If $t \in \mathbb{N}$ is a random variable with $\Pr[t > \eta g(n)] \leq n^{c_1 - \eta c_2}$ for any $\eta \geq \eta'$, then the expected value of t is $O(g(n))$.*

Proof. Let $\eta := \kappa n^i$ for some constant $\kappa \geq \max\{\frac{|c_1|+1}{c_2}, \eta'\}$ and $i \in \mathbb{N}$. Then,

$$\begin{aligned} \Pr[\kappa n^{i+1}g(n) \geq t > \kappa n^i g(n)] &\leq \Pr[t > \kappa n^i g(n)] \\ &\leq n^{c_1 - c_2 \kappa n^i} \\ &\leq n^{c_1 - (|c_1|+1) n^i} \\ &\leq n^{-n^i}. \end{aligned}$$

For $n \geq 2$, the expected value $\mathbb{E}[t]$ is thus

$$\begin{aligned} \mathbb{E}[t] &= \sum_{t'=1}^{\infty} t' \cdot \Pr[t = t'] \\ &\leq \kappa g(n) + \sum_{i=0}^{\infty} \sum_{t' = \kappa n^i g(n)+1}^{\kappa n^{i+1} g(n)} t' \cdot \Pr[t = t'] \\ &\leq \kappa g(n) + \sum_{i=0}^{\infty} \kappa n^{i+1} g(n) \cdot n^{-n^i} \\ &= \kappa g(n)(1 + \sum_{i=0}^{\infty} n^{i+1-n^i}) \\ &\leq \kappa g(n)(2 + \sum_{i=1}^{\infty} n^{-n^i}) \\ &\leq \kappa g(n)(2+2). \end{aligned}$$

$\square$

From this, we can easily derive the following lemma.

Lemma 3.3. *Let $c > 0$, $\varepsilon > 0$, $\eta' > 0$ be constants, let $n \in \mathbb{N}^+$, and let $g(n) > 0$ be a function. If $t \in \mathbb{N}$ is a random variable with $\Pr[t > \eta g(n)] \leq 2^{-\eta c n^{\varepsilon}}$ for any $\eta \geq \eta'$, then the expected value of t is $O(g(n))$.*

Proof. For any $n \in \mathbb{N}^+$ and any constant $\varepsilon > 0$ there exists a positive constant k with $\frac{n^{\varepsilon}}{\log_2 n} \geq k$. Then we have

$$\begin{aligned} 2^{-\eta c n^{\varepsilon}} &= 2^{-\eta c n^{\varepsilon} \frac{\log_2 n}{\log_2 n}} \\ &= n^{-\eta c \frac{n^{\varepsilon}}{\log_2 n}} \\ &\leq n^{-\eta c k}. \end{aligned}$$

Thus, we can apply Lemma 3.2 with $c_1 = 0$ and $c_2 = ck$ to prove the claim. $\square$

Chapter 4
The Single Source Shortest Path Problem

This Chapter is based on the paper "A Tight Analysis of the $(1+1)$-EA for the Single Source Shortest Path Problem" by Benjamin Doerr, E. H., and Christian Klein [DHK07].

4.1 Introduction

The first work in which evolutionary algorithms for classical combinatorial problems were analyzed considered the Single Source Shortest Path problem and the Sorting problem [STW04]. Since already the analysis of simple evolutionary algorithms for simple pseudo-boolean functions were quite technical and complicated, it is not surprising that the analysis of the Single Source Shortest Path problem in [STW04] is tight only for certain instances.

4.1.1 Related Work

In [STW04], Scharnow, Tinnefeld, and Wegener propose a natural $(1+1)$ evolutionary algorithm for the problem of finding shortest paths from a single vertex (the "source") to all other vertices in a graph with edge weights (see below for a precise definition of the problem). They show an upper bound of $O(n^3)$ for the expected optimization time on n-vertex graphs. This bound is tight if (and only if, as we shall see) the graph and edge weights are such that there is a vertex such that all shortest paths to the source contain $\Omega(n)$ edges.

The proof given by Scharnow, Tinnefeld, and Wegener in [STW04] reveals an in fact stronger upper bound of $O(n^2 \sum_{i=1}^n \log(n_i))$, where n_i is the number of vertices for which the shortest path (with respect to the weights) to the source with the minimum number of edges consists of exactly i edges. In particular, since $\sum_{i=1}^n n_i = n - 1$ and thus $\sum_{i=1}^n \log(n_i) \leq \max_{j=1}^{\ell} \{j \log(\frac{n-1}{j})\}$, this yields a bound of $O(n^2 \ell \log(\frac{n}{\ell}))$, where ℓ is the smallest integer such that any vertex can be reached from the source via a shortest path having at most ℓ edges.

4.1.2 Our Results

In this chapter, we give a tight analysis of the (1 + 1) evolutionary algorithm proposed in [STW04]. This leads to an improved upper bound for the expected optimization time of $O(n^2 \max\{\log(n), \ell\})$. In addition, this bound not only holds in expectation, but is fulfilled with high probability, that is, with probability $1 - O(n^{-c})$ for an arbitrary constant c. The bound on the optimization time is tight for all ℓ. For all values of ℓ we present a problem instance such that all shortest paths have length at most ℓ, but the optimization time is $\Omega(n^2 \max\{\log(n), \ell\})$ with high probability.

To prove the upper bound, we develop a method that might see further applications in the future. We closely analyze how nodes become connected to the source via shortest paths. The growth of such shortest paths (note that they do not have to be unique) displays a strong concentration behavior. Although we use a union bound argument over all paths needed, it is still strong enough to obtain bounds that hold with high probability. To show the lower bound, we use the Chernoff type strong concentration bound introduced in Section 3.1.

4.2 An Evolutionary Algorithm for the SSSP Problem

Let $G = (V, E)$ with $V = [1..n]$, $E \subseteq V^2$ be a directed graph with edge weights $w \colon E \to \mathbb{N}$. Given a vertex $s \in V$ called "source", the Single Source Shortest Path (SSSP) problem is the problem of finding a shortest path from s to all other vertices $v \in V \setminus \{s\}$. A path from u to v is a sequence $u = v_0, \ldots, v_k = v$ of vertices such that $(v_{i-1}, v_i) \in E$ for all $i \in [1..k]$ and $v_i \neq v_j$ for $i, j \in [0..k]$, $i \neq j$. The length of a path is the sum of the

4.2 An Evolutionary Algorithm for the SSSP Problem

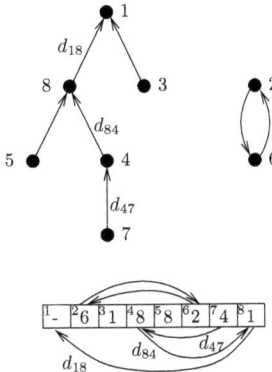

Figure 4.1: The graph representation and the vector representation of an individual not representing a tree. The fitness of vertex 7 is $d_{18}+d_{84}+d_{47}$, as indicated by the edge weights in the graph representation and next to the arrows below the vector representation. The fitness of vertex 2 is ∞, since there is no path from vertex 2 to vertex 1, as shown by the arrows above the vector representation.

weights of the edges it traverses. Dijkstra's famous algorithm [Dij59] solves the problem in time $O(n^2)$.

If we allow the edge weight $w(e) = \infty$ for not existing edges $e \notin E$ we can consider the complete graph K_n on n vertices. The problem instance is given by the distance matrix $D = (d_{ij})_{1 \le i,j \le n}$ of the graph, where $d_{ij} = w((i,j)) \in \mathbb{N} \cup \{\infty\}$. Note that the below described algorithm also works in the case of undirected graphs. For each undirected edge $e = \{i, j\}$ simply set $d_{ij} = d_{ji} = w(e)$.

In this chapter, we analyze the $(1+1)$ evolutionary algorithm (from now on called $(1+1)$-EA$_{sssp}$) for the SSSP problem introduced in [STW04]. We describe and analyze this algorithm assuming that the source is $s = 1$.

4.2.1 Individuals

It is easy to see that we can choose shortest paths from s to any other vertex $i \in V \setminus \{s\}$ in such a way that the union of these paths forms a tree. Hence we may represent solutions to the SSSP problem by giving for each vertex $i \in V \setminus \{s\}$ its predecessor $p(i)$ on a shortest path from s to i. Thus, the candidate solutions can be represented as vectors of predecessors

$I = (p(2), \ldots, p(n)) \in [1..n]^{n-1}$. Note that this representation does not assure that an individual forms a tree. See Figure 4.1 for an example.

4.2.2 Fitness Function

The $(1+1)$-EA$_{sssp}$ uses a multi-criteria fitness function $f \colon [1..n]^{n-1} \to \mathbb{N} \cup \{\infty\}$. For an individual I, it is defined by $f(I) := (f_2(I), \ldots, f_n(I))$ with

$$f_i(I) := \begin{cases} \infty & \text{if } I \text{ does not connect } s \text{ to } i, \\ w(P(s,i)) & \text{otherwise.} \end{cases}$$

Here, $w(P(s,i))$ is the cost of the path P from s to i implied by I. If this path is $P = (s = v_1, v_2, \ldots, v_j = i)$ for $v_1, \ldots, v_j \in V$ then $w(P(s,i)) = d_{v_1 v_2} + \cdots + d_{v_{j-1} v_j}$. See Figure 4.1 for an example. When comparing the fitness values of two individuals I and I', we use $f(I') \leq f(I)$ if $f_i(I') \leq f_i(I)$ for all $2 \leq i \leq n$.

4.2.3 Mutation Operator

As we have explained in Section 2.4, to simulated the behavior of the "classical" $(1+1)$-EA on bit-strings, the mutation step should perform a number of elementary mutations that locally change the individual. The number $S+1$ of elementary mutations performed is distributed according to a Poisson distribution Pois($\zeta = 1$) with parameter $\zeta = 1$. Thus, the probability that in a mutation step $S+1 = k+1$ elementary mutations are performed is $\Pr[S=k] = \frac{1}{ek!}$.

An elementary mutation of the vector I consists of randomly choosing a vertex v with $v \in V \setminus \{s\}$ and setting its predecessor $p(v)$ to a vertex w chosen uniformly at random with $w \in V \setminus \{v\}$. Obviously, there are $(n-1)^2$ possible ways to choose a vertex and its predecessor and thus to do an elementary mutation on individual I.

4.2.4 The $(1+1)$-EA$_{sssp}$

At the beginning, the $(1+1)$-EA$_{sssp}$ generates the initial population $\mathcal{I}$ consisting of an initial individual I. I is created by assigning to each vertex $v \in V \setminus \{s\}$ a predecessor $p(v) \in V \setminus \{v\}$ uniformly at random. In the following mutation step, I is modified to generate a new individual I'. Then, a selection step is done replacing the individual I by I' if the fitness of I' is not

4.3 Upper Bound on the Optimization Time

$(1 + 1)$-EA FOR SSSP

    Initialization:
1    $\mathcal{I} \leftarrow \{I\} = \{(p(2), \ldots, p(n))\}$,
        $p(i) \in V \setminus \{i\}$ chosen u. a. r.
2   **repeat**
        Mutation:
3   Pick S according to $\Pr[S = k] = \frac{1}{e \cdot k!}$
4        $I^0 \leftarrow I$
5   **for** $m = 1$ to $S + 1$
6   **do**
7   Choose $i \in [2..n]$ u. a. r.
8   Choose $j \in [1..n] \setminus \{i\}$ u. a. r.
9   Generate I^m from I^{m-1} by setting $p(i)$ to j.
        Selection:
10   **if** $f_i(I^{S+1}) \leq f_i(I)$ for all $i \in [2..n]$
11   **then** $\mathcal{I} \leftarrow \{I^{S+1}\}$
12   **until** $\mathcal{I}$ contains an optimal solution

Figure 4.2: Pseudocode for the $(1+1)$-EA$_{sssp}$ for the SSSP problem.

worse than I's fitness. Mutation and selection are repeated until an optimal solution is found. Pseudocode for the $(1+1)$-EA$_{sssp}$ for the SSSP problem is given in Figure 4.2.

Note that the selection step accepts a new individual I' only if $f(I') \leq f(I)$ which is the case if $f_i(I') \leq f_i(I)$ for all $2 \leq i \leq n$. That means that for every vertex $i \in [2..n]$ the path from s to i in I' is at most as long as the path from s to i in I. Therefore, once we have found an optimal path for a vertex v, the $(1+1)$-EA$_{sssp}$ does not accept mutations that would cause s to be connected to v using a suboptimal path.

4.3 Upper Bound on the Optimization Time

In this section we show that the optimization time of the $(1+1)$-EA$_{sssp}$ for the SSSP problem is $O(n^2 \max\{\log(n), \ell\})$ with high probability. Here, ℓ is the maximum number of edges of all shortest paths with a minimum number of edges (see definition below). Recall that with high probability means that an event happens with probability at least $1-O(n^{-c})$, where c is an arbitrary constant.

The key observation in the proof is that the actual growth of the shortest path tree deviates only little from its expected growth. This phenomenon is called *strong concentration* and can be quantized through so-called Chernoff bounds as described in Section 3.1.

We will need the following definition.

Definition 4.1 (Edge Radius). *The edge radius $\ell_G(u)$ of a vertex u in a weighted graph G is the maximum number of edges of a shortest path with minimum number of edges from u to a vertex v in G. That is*

$$\ell_G(u) = \max_{v \in V} \min_{P \in \mathcal{P}_{uv}} \ell(P)$$

with $\mathcal{P}_{uv} := \{P \mid P \text{ is a shortest path from } u \text{ to } v\}$ and $\ell(P)$ being the number of edges of path P.

For the rest of this chapter, we will denote by $\ell := \ell_G(s)$ the edge radius of the source s in G.

Now we can prove the upper bound.

Theorem 4.1. *The optimization time of the $(1+1)$-EA_{sssp} is with high probability $O(n^2 \max\{\log(n), \ell\})$.*

This follows immediately from the following lemma.

Lemma 4.1. *Let $\ell^* := \max\{\log(n), \ell\}$, $c > 0$ be a constant, $\eta \geq 8(c+1) > 8$ and $t := e\eta\ell^*(n-1)^2$. Then the optimization time needed by the $(1+1)$-EA_{sssp} to find all shortest paths is less than t with probability $p > 1 - n^{-c}$.*

Proof. Because of the multi-criteria fitness function, the $(1+1)$-EA_{sssp} cannot replace any path in the individual I by a longer path. Thus, in the result of a successful mutation step all paths are either the same as before or are replaced by a path that is not longer. Hence, any successful mutation step that would apply more than one elementary mutation can be simulated by a number of successful mutation steps applying a single elementary mutation. Since the probability for a mutation step consisting of a single elementary mutation is $\frac{1}{e}$ and thus constant, for the upper bound analysis we can assume that only mutation steps performing a single elementary mutation are successfully applied.

4.3 Upper Bound on the Optimization Time

The $(1+1)$-EA$_{sssp}$ has to find $n-1$ shortest paths from the source s to all other vertices. Note that there may be many different possible shortest paths for a vertex v.

Pick a vertex v and a shortest path $P := (v_1, \ldots, v_{\ell'+1})$ from $s = v_1$ to $v = v_{\ell'+1}$. Note that by the definition of the edge radius ℓ we can pick P so that it has $\ell' \leq \ell$ edges. We call a mutation step performing a single elementary mutation the j-th improvement in P if prior to the mutation the individual I contains a shortest path from s to v_j for some $1 \leq j \leq \ell'$ and after the mutation step the predecessor of v_{j+1} is $p(v_{j+1}) = v_j$. Note that after the j-th improvement I contains a shortest path from s to v_{j+1}, but that it might have already contained such a shortest path before. If the $(1+1)$-EA$_{sssp}$ has performed the ℓ'-th improvement we say it has *followed* P. Obviously, a shortest path from s to $v = v_{\ell'+1}$ has been found by then.

Let $t := e\eta\ell^*(n-1)^2$ and let t' be the number of steps the $(1+1)$-EA$_{sssp}$ needs to follow P. Define the random variables X_i for $1 \leq i \leq t'$ by $X_i = 1$ if the i-th mutation step is an improvement in P and $X_i = 0$ otherwise. Then independent of the $i-1$ steps before, $\Pr[X_i = 1] \geq p := \frac{1}{e}\frac{1}{(n-1)^2}$, since either the corresponding predecessor is already set correctly before or the i-th mutation step consists of a single elementary mutation (with probability $\frac{1}{e}$), picks the correct vertex (with probability $\frac{1}{n-1}$), and sets it to its correct predecessor (with probability $\frac{1}{n-1}$). For $t' < i \leq t$ define the random variables X_i by $\Pr[X_i = 1] := p$ and $\Pr[X_i = 0] := 1-p$. Obviously, X_i is independent of X_j for all $1 \leq i < j \leq t$.

Now, let X_i^* for $i \in [1..t]$ be mutually independent random binary variables with $\Pr[X_i^* = 1] := p$ and $\Pr[X_i^* = 0] := 1-p$ and let $X^* := \sum_{i=1}^{t} X_i^*$. The expected value of $X := \sum_{i=1}^{t} X_i$ is

$$\mathbb{E}[X] \geq \mathbb{E}[X^*] = pt = \frac{1}{e}\frac{1}{(n-1)^2}e\eta\ell^*(n-1)^2 = \eta\ell^*.$$

If the $(1+1)$-EA$_{sssp}$ has not found an optimal path from s to v, it obviously has not followed P and thus $X < \ell$. Since $\Pr[X_i = 1 \mid X_1 = x_1, \ldots, X_{i-1} = x_{i-1}] \geq \Pr[X_i^* = 1]$ for all i and all $x_1, \ldots, x_{i-1} \in \{0,1\}$, we can apply Lemma 3.1 to bound the probability of not finding a shortest path from s to v in time t by

$$\begin{aligned}
\Pr\begin{bmatrix}\text{no shortest path from}\\ s \text{ to } v \text{ found in time } t\end{bmatrix} &\leq \Pr\begin{bmatrix}P \text{ not followed}\\ \text{in time } t\end{bmatrix} \\
&\leq \Pr[X < \ell] \\
&\leq \Pr[X < \ell^*] \\
&\leq \Pr[X^* < \ell^*].
\end{aligned}$$

Using $\alpha := \frac{1}{\eta} < 1$, we can use the first inequality of Theorem 3.1 to bound the probability of not finding a shortest path from s to v in time $t = e\eta\ell^*(n-1)^2$ by

$$\begin{aligned}
\Pr\begin{bmatrix}\text{no shortest path from}\\ s \text{ to } v \text{ found in time } t\end{bmatrix} &\leq \Pr[X^* < \ell^*]\\
&= \Pr[X^* < \alpha\mathbb{E}[X^*]]\\
&\leq \exp(-\frac{1}{2}(1-\alpha)^2\mathbb{E}[X^*])\\
&= \exp\left(-\frac{1}{2}\frac{(\eta-1)^2}{\eta^2}\eta\ell^*\right)\\
&\leq \exp\left(-\frac{(\frac{\eta}{2})^2\ell^*}{2\eta}\right)\\
&= \exp\left(-\frac{\eta}{8}\ell^*\right)
\end{aligned}$$

since $\eta > 8$.

Using a union bound argument we get

$$\begin{aligned}
\Pr\begin{bmatrix}\text{not for all } v_i \text{ a shortest path}\\ \text{from } s \text{ to } v_i \text{ found in time } t\end{bmatrix} &\leq \sum_{i=2}^{n}\Pr\begin{bmatrix}\text{no shortest path from } s\\ \text{to } v_i \text{ found in time } t\end{bmatrix}\\
&\leq \sum_{i=2}^{n}\exp\left(-\frac{\eta}{8}\ell^*\right)\\
&\leq n\exp\left(-\frac{\eta}{8}\log(n)\right)\\
&\leq n^{1-\frac{\eta}{8}}\\
&\leq n^{-c}.
\end{aligned}$$

$\square$

Note that we did not optimize for η.

From the fact that the upper bound of $O(n^2\max\{\log(n),\ell\})$ holds with high probability, we can derive an upper bound on the expected optimization time.

Theorem 4.2. *The* $(1+1)$-EA_{sssp} *has an expected optimization time of* $O(n^2\max\{\log(n),\ell\})$.

Proof. This follows directly from Lemma 4.1 in combination with Lemma 3.2.
$\square$

4.4 Lower Bound on the Optimization Time

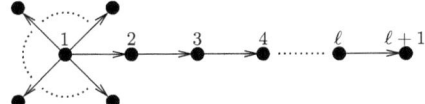

Figure 4.3: The SSSP tree of $G_{n,\ell}$ with source $s = 1$.

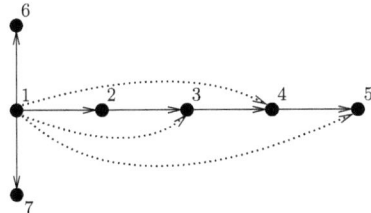

Figure 4.4: The weighted graph $G_{7,4}$. The solid edges have weight one and form the shortest path tree. All other edges have weight 14. The dotted edges are the second shortest path from 1 to i for $i \in \{3, 4, 5\}$.

4.4 Lower Bound on the Optimization Time

In this section we show a lower bound matching the upper bound presented in the previous section. More precisely, for any $n \in \mathbb{N}$ and $\ell \in [1..n-1]$ we define a weighted graph $G_{n,\ell}$ on n vertices with edge radius $\ell_{G_{n,\ell}}(s) = \ell$ for which the algorithm has an optimization time of at least $\Omega(n^2 \max\{\log(n), \ell\})$ with high probability.

4.4.1 A Worst Case Graph Class

Let $n \in \mathbb{N}$, $V = [1..n]$. For all $\ell \in [1..n-1]$ we define the weighted graph $G_{n,\ell} = (V, E)$ such that the source of the SSSP tree to be computed is $s = 1$ and $\ell_{G_{n,\ell}}(s) = \ell$. We show that the optimization time of the $(1+1)$-EA$_{sssp}$ is $\Omega(n^2 \max\{\log(n), \ell\})$ with high probability.

We will set the weights in such a way that $(1, 2, \ldots, \ell, \ell+1)$ is the unique shortest path from $s = 1$ to $\ell + 1$. For all other vertices k with $k > \ell + 1$, the edge (s, k) shall be the unique shortest path from s to k. Figure 4.3 shows the SSSP tree of $G_{n,\ell}$. For simplicity, we assign the weight 1 to all edges in the SSSP tree.

To guarantee that the optimization time depends linearly on ℓ, the edges on the shortest path should be added one by one. To this purpose, each vertex $i \in [3..\ell+1]$ should be connected to vertex s with a sufficiently cheap edge ensuring that, as long as vertex $i-1$ is not correctly connected, it is cheaper to connect s and i directly than to connect s to i via $i-1$. These requirements are fulfilled by $G_{n,\ell} := ([1..n], \{(i,j) \mid u,v \in [1..n], u \neq v\})$ with edge weights

$$w(i,j) := \begin{cases} 1, & \text{if } j = i+1 \leq \ell+1, \\ 1, & \text{if } i = 1 \wedge j > \ell+1, \\ 2n, & \text{otherwise.} \end{cases}$$

The graph $G_{7,4}$ is shown in Figure 4.4. Note that $G_{n,1}$ is the graph with edge weight 1 for each edge $(s,i), i \in [2..n]$ and $2n$ for all other edges.

4.4.2 A Lower Bound

We now give a lower bound depending on n and ℓ on the number of steps needed by the $(1+1)$-EA$_{sssp}$ to find an optimal solution. To prove that $\Omega(n^2 \max\{\log(n), \ell\})$ is a lower bound on the optimization time, we first prove that $\Omega(n^2 \log(n))$ is a lower bound on the optimization time of the $(1+1)$-EA$_{sssp}$. Observe that this bound holds for all graphs that have a unique SSSP tree.

Lemma 4.2. *Let $G = (V, E)$ be a graph on n vertices and $s \in V$ a vertex such that the SSSP tree of G with source s is unique. Then the number of steps needed by the $(1+1)$-EA$_{sssp}$ to compute the SSSP tree of G with source s is $\Omega(n^2 \log(n))$ with high probability.*

Proof. To proof this bound our approach is as follows. First we bound the probability that a fixed elementary mutation connecting one vertex to its predecessor is never tried during $t-1$ steps. From this we bound the probability that at least one of the vertices is not connected to its predecessor after $t-1$ steps. This is the probability that the $(1+1)$-EA$_{sssp}$ needs at least t steps to connect all vertices to their predecessors and thus to find an optimal solution.

Some arguments of this proof are similar to the proof for the coupon collector's problem (cf. the book by Motwani and Raghavan [MR95]).

Due to the uniqueness of the SSSP tree and the fact that each vertex of G is incident with $n-1$ edges, the probability that a vertex is connected correctly after the initialization step is $\Theta(n^{-1})$. Thus, the probability that

4.4 Lower Bound on the Optimization Time

c vertices are correctly initialized is $\Theta(n^{-c})$. Since we know that the upper bound on the optimization time is $O(n^2 \max\{\ell, \log(n)\})$ we get that with high probability c is constant. Thus, with high probability $n-1-c$ vertices remain that are not connected correctly after the initialization. For these vertices, the $(1+1)$-EA$_{sssp}$ has to perform the corresponding $n-1-c$ elementary mutations that connect them to their correct predecessor.

Let T denote the random variable describing the number of steps needed by the $(1+1)$-EA$_{sssp}$ to find the shortest path tree with source s. We now calculate the probability $\Pr[T \geq t]$ that the $(1+1)$-EA$_{sssp}$ needs at least t steps. This is the same as the probability that the $(1+1)$-EA$_{sssp}$ will not finish after $t-1$ steps, meaning that at least one of the $n-1-c$ remaining vertices are not connected to their correct predecessor during these $t-1$ steps.

Consider one of the $n-1-c$ vertices not yet connected correctly. During the construction of the shortest path tree, this vertex v has to be chosen at least once together with its correct predecessor v' to connect it using the right edge. The probability that this happens in a single elementary mutation is $\frac{1}{n-1} \cdot \frac{1}{n-1} = \frac{1}{(n-1)^2}$.

By the definition of the Poisson distribution, the probability that during a single step exactly $S+1$ elementary mutations are performed is $\frac{1}{eS!}$. Since each of the $S+1$ elementary mutations in this step are done independently, the probability that a fixed elementary mutation is performed in this step consisting of $S+1$ elementary mutations is at most $\frac{S+1}{(n-1)^2}$.

Hence, the probability that a fixed elementary mutation is performed in one step is at most

$$\sum_{S=0}^{\infty} \frac{1}{e \cdot S!} \cdot \frac{S+1}{(n-1)^2} = \frac{2}{(n-1)^2}$$

since

$$\sum_{S=0}^{\infty} \frac{S+1}{S!} = \sum_{S=1}^{\infty} \frac{S}{S!} + \sum_{S=0}^{\infty} \frac{1}{S!}$$
$$= \sum_{S=1}^{\infty} \frac{1}{(S-1)!} + e$$
$$= 2e.$$

By this, the probability that a fixed elementary mutation is never chosen during $t-1$ steps is at least $(1 - \frac{2}{(n-1)^2})^{t-1}$. Hence it follows that this fixed elementary mutation is tried at least once in $t.-1$ steps with probability at most $1-(1-\frac{2}{(n-1)^2})^{t-1}$.

There are $n-1-c$ correct elementary mutations that have to be performed. Denote by E_i for $i \in [1..n-1-c]$ the event that the i-th of the $n-1-c$ vertices is connected to the correct predecessor within $t-1$ steps. The order in which they are connected does not matter. By the above calculations $\Pr[E_1] \leq 1-(1-\frac{2}{(n-1)^2})^{t-1}$.

Note that the events E_i are negatively correlated, meaning that if one event E_i is true, the probability for another E_j decreases. The reason for this is the following: Since one of the elementary mutations performed connects the i-th vertex to its correct predecessor, there are less elementary mutations remaining that can possibly connect the j-th vertex to its correct predecessor. Hence, the probability that all events E_i take place, that is all necessary $n-1-c$ elementary mutations are performed within $t-1$ steps, is bounded by

$$\begin{aligned}\Pr[E_1,\ldots,E_{n-1-c}] &= \prod_{i=1}^{n-1-c} \Pr[E_i \mid E_1,\ldots,E_{i-1}] \\ &< \prod_{i=1}^{n-1-c} \Pr[E_1] \\ &\leq (1-(1-\frac{2}{(n-1)^2})^{t-1})^{n-1-c}.\end{aligned}$$

But this means that with probability at least $1-(1-(1-\frac{2}{(n-1)^2})^{t-1})^{n-1-c}$ at least one of the combinations of a vertex and its correct predecessor is never chosen during the $t-1$ steps. This is the probability $\Pr[T \geq t]$ that the $(1+1)$-EA_{sssp} needs at least t steps.

We now choose $t := 1+(\frac{1}{2}(n-1)^2-1)\frac{1}{2}\log(n-1-c) = \Omega(n^2 \log(n))$. Using the inequality $(1-\frac{1}{k})^k < e^{-1} < (1-\frac{1}{k})^{k-1}$ we obtain the following bound for the probability $\Pr[T \geq t]$ that not all $n-1-c$ vertices are connected to their predecessors within $t-1$ steps.

4.4 Lower Bound on the Optimization Time 31

$$\begin{aligned}
\Pr[T \geq t] &\geq 1 - \left(1 - \left(1 - \frac{2}{(n-1)^2}\right)^{(\frac{1}{2}(n-1)^2 - 1)\frac{1}{2}\log(n-1-c)}\right)^{n-1-c} \\
&= 1 - \left(1 - \left(\left(1 - \frac{2}{(n-1)^2}\right)^{\frac{1}{2}(n-1)^2 - 1}\right)^{\frac{1}{2}\log(n-1-c)}\right)^{n-1-c} \\
&\geq 1 - \left(1 - \left(e^{-1}\right)^{\frac{1}{2}\log(n-1-c)}\right)^{n-1-c} \\
&= 1 - \left(1 - \frac{1}{\sqrt{n-1-c}}\right)^{n-1-c} \\
&= 1 - \left(\left(1 - \frac{1}{\sqrt{n-1-c}}\right)^{\sqrt{n-1-c}}\right)^{\sqrt{n-1-c}} \\
&\geq 1 - e^{-\sqrt{n-1-c}}.
\end{aligned}$$

Hence the $(1 + 1)$-EA$_{sssp}$ will need at least $t = \Omega(n^2 \log(n))$ steps with high probability. □

By the above lemma, our lower bound is tight as long as $\ell \in O(\log(n))$. To complete our claim, however, we need to prove that for larger ℓ the optimization time linearly depends on ℓ.

Lemma 4.3. *Let $n \in \mathbb{N}$ and $\ell \in \omega(\log(n))$. Then the optimization time of the $(1 + 1)$-EA$_{sssp}$ on $G_{n,\ell}$ is $\Omega(n^2 \ell)$ with high probability.*

Proof. The idea of this proof is similar to the one used in [DJW02] for the proof of the lower bound on the runtime of the $(1+1)$-EA on leading ones (LO).

To prove the claim, we analyze how long it takes until the individual I contains the path $P := (s = 1, 2, \ldots, \ell, \ell+1)$. To this aim, we analyze how the length $L(I)$ of the longest subpath of P starting with s that is contained in I grows. Note that this length $L(I)$ never decreases, since for each vertex on P this subpath is the unique shortest path to s. In the following, we show that with high probability (i) $L(I)$ initially is constant, (ii) in $\Theta(n^2\ell)$ iterations $L(I)$ increases at most $O(\ell)$ times, and (iii) the total increase in these $O(\ell)$ relevant iterations (plus the initial constant length) is less than ℓ.

The probability that in the initial individual some vertex $i \in [2..\ell+1]$ is already linked to $i-1$ is exactly $\frac{1}{n-1}$. Hence the probability that $L(I) \geq c$ is $O(n^{-c})$ for all $c \in \mathbb{N}$, that is, with high probability $L(I)$ is initially constant.

Let $t := \eta(n-1)^2\ell$ and let t^* be the time step when $L(I)$ increases for the $4\eta\ell$-th time for a constant η which we will determine later on. For $i \in [1..t^*]$, we define a binary random variable X_i by $X_i = 1$ if $L(I)$ increases in step i. To increase $L(I)$, one of the $S+1$ elementary mutations in the current mutation step has to connect vertex $L(I)+2$ to vertex $L(I)+1$ and this mutation step to be accepted. The probability that an elementary mutation succeeds in connecting $L(I)+2$ to $L(I)+1$ is obviously $\frac{1}{(n-1)^2}$. Hence, as in the proof of Lemma 4.2, the probability that one mutation step does so is at most $\frac{2}{(n-1)^2}$ and hence $\Pr[X_i = 1] \leq p := \frac{2}{(n-1)^2}$. For $i \in [t^*+1..t]$ define X_i by $\Pr[X_i = 1] := p$ and $\Pr[X_i = 0] := 1-p$. Clearly, X_i is independent of X_j for $1 \leq i < i \leq t$.

Now define X_i^* for $i \in [1..t]$ to be mutually independent random binary variables with $\Pr[X_i^* = 1] := p$ and $\Pr[X_i^* = 0] := 1-p$ for all i and let $X^* := \sum_{i=1}^{t} X_i^*$. Then $\Pr[X_i^* = 1] \geq \Pr[X_i = 1 \mid X_1 = x_1, \ldots X_{i-1} = x_{i-1}]$ for all $x_1, \ldots, x_{i-1} \in \{0,1\}$. Then the expected value of $X := \sum_{i=1}^{t} X_i$ is

$$\mathbb{E}[X] \leq \mathbb{E}[X^*] = \eta(n-1)^2 \ell \frac{2}{(n-1)^2} = 2\eta\ell.$$

Hence, by applying Lemma 3.1 and using the third Chernoff bound from Theorem 3.1 with $\gamma = 1$ we get

$$\begin{aligned}
\Pr[X \geq 4\eta\ell] &\leq \Pr[X^* \geq 2\mathbb{E}[X^*]] \\
&\leq \exp\left(-\frac{\mathbb{E}[X]}{3}\right) \\
&= \exp\left(-\frac{2\eta\ell}{3}\right) \\
&= \exp\left(-\frac{2\eta}{3}\log(n)\frac{\ell}{\log(n)}\right) \\
&= n^{-\frac{2\eta}{3}\frac{\ell}{\log(n)}} \\
&= n^{-\omega(1)}.
\end{aligned}$$

In the last lines we used that since $\ell \in \omega(\log(n))$ we have $\frac{\ell}{\log n} = \omega(1)$ for any constant η. But this means, that with high probability the $(1+1)$-EA does at most $t' := 4\eta\ell$ improvements during the t steps.

Finally, we analyze how many additional vertices might become connected to s via the shortest path during the t' improvements. For that, we analyze the additional amount by which such an improvement increases $L(I)$. To this end, note that there are two ways how an additional vertex i can be connected to the longest subpath of P starting in s. Either (i) there may be

4.4 Lower Bound on the Optimization Time

an elementary mutation in the mutation step that causes the improvement that changes the pointer of i to its predecessor $i-1$ in P, or (ii), i may coincidentally be connected to $i-1$, which becomes part of the subpath by an event of type (i) or (ii). We shall argue that both events happen only with a probability of at most $\frac{1}{2}$.

(i) Suppose first that i enlarges the path of interest through an elementary mutation. For this to happen (among other things), the following has to occur. Among the possible more than one elementary mutations in the current step that connect i to some other vertex, the last one has to connect i to its predecessor $i-1$ in P. By definition of the mutation operator, this happens with a probability of $\frac{1}{n-1} \leq \frac{1}{2}$ for $n \geq 3$.

(ii) Now suppose that the predecessor $i-1$ of i becomes part of the subpath of interest. We argue that the probability that i is coincidentally pointing to $i-1$ is at most the probability that it is pointing to s, and in consequence, at most $\frac{1}{2}$. There are two possible reasons why i might point to $i-1$, either (a) in the initialization i's predecessor is set to $i-1$ or (b) i's predecessor is set to $i-1$ in some mutation step.

(a) For the initially chosen individual the probability that i points to $i-1$ is clearly as big as the probability that it points to s, since the predecessor if i is chosen uniformly at random from all vertices different from i.

(b) Consider an iteration that does not result in making $i-1$ part of the subpath of interest, and fix a sequence of elementary mutations to be conducted in this iteration. Assume that at the start of the iteration i's predecessor is vertex $j \in [1..n] \setminus \{i\}$, for which the path from s to j in I has length $w_-(s,j)$, and that at the end of the iteration the length of the path from s to $i-1$ in I is $w_+(s,i-1)$. If $w_+(s,i-1) \leq w_-(s,j)$, the acceptance of the iteration does not depend on whether i's pointer is changed to $i-1$ or to s, since both events would not increase the length of the shortest path from i to s in I. On the other hand, if $w_+(s,i-1) > w_-(s,j)$, only a change of i's pointer to s, but not to $i-1$, would be accepted. Thus, any iteration that does not make $i-1$ part of the path rather increases the probability that i points to s compared to the probability of pointing to $i-1$. In consequence, the probability that i coincidentally points at $i-1$ in the iteration in which $i-1$ becomes part of the subpath of interest, is at most $\frac{1}{2}$.

Summarizing, an additional vertex may either already be coincidentally connected correctly (with probability at most $\frac{1}{2}$), or, if this is not the case (which happens with probability at least $\frac{1}{2}$), it may become connected by an elementary mutation (with probability at most $\frac{1}{2}$). Hence, the probability that an additional vertex becomes connected is at most $\frac{3}{4}$.

Chapter 4: The Single Source Shortest Path Problem

Let t'' be the number of improvement steps the $(1+1)$-EA$_{sssp}$ performs until it finds the optimal solution. For $i \in [1..t'']$ let Y_i be the random variable describing the number of additional vertices that are added in the i-th improvement step. By the above arguments, independent of the outcome of the random choice of S, $\Pr[Y_i = m] \leq (\frac{3}{4})^{m-1}\frac{1}{4}$. If $t'' < t' = 4\eta\ell$ let Y_i for $i \in [t''+1..t']$ be defined by $\Pr[Y_i = m] := (\frac{3}{4})^{m-1}\frac{1}{4}$. Obviously, Y_i is independent of all Y_j for $i < j$. Define Y_i^* for $i \in [1..t']$ to be mutually independent random variables that are geometrically distributed with parameter $q = \frac{1}{4}$, that is, $\Pr[Y_i^* = m] := (\frac{3}{4})^{m-1}\frac{1}{4}$ for all $m \in \mathbb{N}$. The expected value of Y_i^* is $\mathbb{E}[Y_i^*] = q^{-1} = 4$. Let $Y := \sum_{i=1}^{t'} Y_i$ and $Y^* := \sum_{i=1}^{t'} Y_i^*$. Then $\mathbb{E}[Y] \leq \mathbb{E}[Y^*] = 4t'$.

Applying Lemma 3.1 and the bound in Theorem 3.2 with $\delta = 1$ and assuming $t' = 4\eta\ell \geq 2$ we get

$$\begin{aligned}
\Pr[Y > 8t'] &\leq \Pr[Y^* > 8t'] \\
&= \Pr[Y^* > 32\eta\ell] \\
&\leq \exp\left(-\frac{(t'-1)}{4}\right) \\
&\leq \exp\left(-\frac{t'}{2 \cdot 4}\right) \\
&= \exp\left(-\frac{\eta\ell}{2}\right) \\
&= \exp\left(-\frac{\eta}{2}\log(n)\frac{\ell}{\log(n)}\right) \\
&= n^{-\frac{\eta}{2}\frac{\ell}{\log(n)}} \\
&= n^{-\omega(1)}.
\end{aligned}$$

Thus, with probability at most $n^{-\omega(1)}$ during up to $t' = 4\eta\ell$ improvements more than $32\eta\ell$ additional vertices become part of the shortest path. Since with high probability the number t' of improvements done in $t = \eta(n-1)^2\ell$ steps is smaller than t', with high probability the number of additional vertices added during t steps is at most $32\eta\ell$.

Now let c be the (constant) number of vertices that are initially correctly linked, X^* is an upper bound on the number of improvements done in $t = \eta(n-1)^2\ell$ steps, and Y^* is with high probability an upper bound on the number of additional vertices discovered during these improvements. Hence with high probability, after t steps, the discovered part of P has length at most $c + X^* + Y^*$. By the above considerations, at most $c + X^* + Y^* \leq$

4.5 Summary

$c+4\eta\ell+32\eta\ell = c+36\eta\ell$ vertices are discovered in time t with high probability. Choosing $\eta := \frac{1}{72}$ concludes the proof for $\ell > 2(c+1)$. □

Combining Lemma 4.2 and Lemma 4.3 yields the following theorem.

Theorem 4.3. *The optimization time needed by the* $(1+1)$-EA_{sssp} *to solve the SSSP problem is* $\Omega(n^2 \max\{\log(n), \ell\})$ *with high probability.*

Observe that this theorem also implies an expected optimization time of $\Omega(n^2 \max\{\log(n), \ell\})$.

4.5 Summary

In this chapter, we gave a tight runtime analysis of the $(1+1)$-EA_{sssp} for the SSSP problem introduced in [STW04]. This includes an improvement of the previous upper bound of $O(n^2 \ell \log(\frac{n}{\ell}))$ (which is implicit in a proof) to $O(n^2 \max\{\ell, \log(n)\})$ and a carefully selected lower bound example for all values of n and ℓ.

At least as important as the precise bounds for this particular problem are the methods developed in this chapter. Past arguments suggested a coupon-collector like behavior in finding the shortest paths. Those, however, cannot be employed to obtain such sharp bounds. Indeed, our analysis shows that the true behavior is different. Namely, the different shortest paths grow at comparable speeds that are strongly concentrated around their expected values.

While our work is very satisfying from the methodological point of view, some particular questions for the SSSP problem remain open. The most challenging one from a broader perspective is whether the multi-criterial fitness function is necessary. Recall that we accept a newly created individual only if for no vertex the distance to the source is increased. A natural (single-criterial) alternative would be to consider the average distance. Scharnow et al. argue in [STW04] that the multi-criterial fitness function is necessary for the algorithm to run properly. However, their counterexample only works if vertices not connected to the source are assumed to have an infinite distance to the source. In this case, changing the number of ∞–distance vertices does not change the average distance, and hence the EA finds itself on a large plateau of constant fitness. A simple way to overcome this (and the one you would choose naturally in an implementation) would be to replace the infinite distance of such vertices by a large, but finite number. We strongly believe that in this case, taking the average distance as fitness would result in the same runtime behavior. However, the analysis seems to be much harder, due to the fact that the individual can develop in less conservative ways.

This problem resembles the problem of maximizing a linear function $g\colon \{0,1\}^n \to \mathbb{R}$, $x \mapsto \sum_{i=1}^n a_i x_i$ with positive coefficients a_i. Employing a $(1+1)$ evolutionary algorithm using the multi-criteria fitness function $f\colon \{0,1\}^n \to \mathbb{R}^n$, $x \mapsto (a_1 x_1, \ldots, a_n x_n)$, a simple coupon collector argument shows an optimization time of $\Theta(n \log(n))$. That the same bound also holds if we use g itself as the fitness function is the result of a highly complex analysis [DJW02]. Attempts to simplify this result later led to the invention of the drift analysis method in evolutionary computation (cf. [HY04]). With this development in mind, it seems likely that it is very difficult to prove that a single-criterial EA can solve the SSSP problem efficiently.

Chapter 5

The All-Pairs Shortest Path Problem

This Chapter is based on the paper "Crossover Can Provably be Useful in Evolutionary Computation" by Benjamin Doerr, E. H., and Christian Klein [DHK08].

5.1 Introduction

The paradigm of nature-inspired computing suggests to use both a mutation and a crossover operator. However, the fundamental question whether crossover is really useful is still not answered in a satisfying way. Most evolutionary algorithms used in practice employ both a mutation operator (that generates a new individual by slightly altering a single parent individual) and a crossover operator (that generates a new individual by recombining information from two parents).

So far, apart from a few artificial examples, no problem was known where an evolutionary algorithm using crossover and mutation is superior to one that only uses mutation.

5.1.1 Related Work

In contrast to the positive practical application of a crossover operator, there is little evidence for the need of crossover. In fact, early work in this direction suggests the opposite. In [MHF93], Mitchell, Holland and Forrest experimentally compared the runtime of a simple genetic algorithm (using crossover) and several hill-climbing heuristics on so-called *royal road functions*. According to Holland's [Hol75] *building block hypothesis*, these functions should be particularly suited to be optimized by an algorithm employing crossover. The experiments conducted in [MHF93], however, clearly demonstrated that this advantage does not exist. In fact, an elementary randomized hill-climbing heuristic (repeated mutation and selection of the fitter one of parent and offspring) was found to be far superior to the genetic algorithm.

The first theoretical analysis indicating that crossover can be useful was given by Jansen and Wegener [JW99] in 1999 (see also [JW02]). For $m < n$, they defined a pseudo-Boolean *jump* function $j_m \colon \{0,1\}^n \to \mathbb{R}$ such that (more or less) $j_m(x)$ is the number of ones in the bit-string x if this is at most $n - m$ or equal to n, but small otherwise. A typical mutation based evolutionary algorithm (flipping each bit independently with probability $1/n$) will easily find an individual x such that $j_m(x) = n - m$, but will need expected time $\Omega(n^m)$ to flip the remaining m bits (all in one mutation step). However, if we add the uniform crossover operator (here, each bit of the offspring is randomly chosen from one of the two parents) and use it sufficiently seldom compared to the mutation operator, then the runtime reduces to $O(n^2 \log n + 2^{2m} n \log n)$. While the precise computations are far from trivial, this behavior stems naturally from the definition of the jump function.

The work of Jansen and Wegener [JW99, JW02] was subsequently extended by different authors in several directions [SW04, JW05], partly to overcome the critique that in the first works the crossover operator necessarily had to be used very sparingly. While these works enlarged the theoretical understanding of different crossover operators, they could not resolve the feeling that all these pseudo-Boolean functions were artificially tailored to demonstrate a particular phenomenon. In [JW05], the authors state that "It will take many major steps to prove rigorously that crossover is essential for typical applications."

The only two works (that we are aware of) that address the use of crossover for other problems than maximizing a pseudo-Boolean function are "Crossover is Provably Essential for the Ising Model on Trees" [Sud05] by Sudholt and "The Ising Model on the Ring: Mutation Versus Recombination" [FW04] by Fischer and Wegener. They show that crossover also helps when considering a simplified Ising model on special graph classes, namely rings and trees. The simplified Ising model, however, is equivalent to looking for a vertex coloring of a graph such that all vertices receive the same color. While it is interesting to see that evolutionary algorithms have difficulties

5.1 Introduction

addressing such problems, proving "rigorously that crossover is essential for typical applications" remains an open problem.

Though not the focus of this theory-driven work, we note that path problems do find significant attention from the evolutionary algorithms community, see ,e. g., [LZHH02, LZHH06, AR02, IHK99].

5.1.2 Our Results

We answer the question of whether crossover is really useful positively. In this chapter, we present the first non-artificial problem for which crossover provably reduces the order of magnitude of the optimization time. This problem is the All-Pairs Shortest Path (APSP) problem, that is, the problem to find, for all pairs of vertices of a directed graph with edge weights, the shortest path from the first vertex to the second. This is one of the most fundamental problems in graph algorithms, see for example the books by Mehlhorn and Näher [MN99] or Cormen et al. [CLRS03].

We present a natural evolutionary algorithm for the APSP problem. It has a population consisting of at most one path for every pair of vertices (connecting the first to the second vertex). Initially, it contains all paths consisting of one edge. A mutation step consists of taking a single path from the population uniformly at random and adding or deleting a (Poisson distributed) random number of times an edge at one of its endpoints. The newly generated individual replaces an existing one (connecting the same vertices) if it is not longer. Hence our fitness function (which is to be minimized) is the length of the path.

We analyze this algorithm and prove that, in the worst case, it has with high probability an optimization time of $\Theta(n^4)$, where n is the number of vertices of the input graph.

We additionally state three different crossover operators for this problem. They all take two random individuals from the population and try to combine them to form a new one. In most cases, of course, this will not generate a path. In this case, we define the fitness of the new individual to be infinite (or some number larger than n times the longest edge). Again, the new individual replaces one having the same endpoints and not smaller fitness.

Using an arbitrary constant crossover rate for any of these crossover operators, we prove an upper bound of $O(n^{3.5+\varepsilon})$ for the expected optimization time. Hence for the APSP problem, crossover leads to a reduction of the optimization time. While the improvement of order $n^{0.5-\varepsilon}$ might not be too important, this work solves a long-standing problem in the theory of evolutionary computation. It justifies to use both a mutation and crossover operator in applications of evolutionary computation.

While our proofs seem to use only simple probabilistic arguments, a closer look reveals that we also invented an interesting tool for the analysis of evolutionary algorithms. A classical problem in the analysis of such algorithms is that the mutation operator may change an individual at several places (multi-bit flips in the bit-string model). Hence unlike for the heuristic of Randomized Local Search (RLS), with evolutionary algorithms we cannot rely on the fact that our offspring is in a close neighborhood of the original search point. While this is intended from the view-point of algorithm design (to prevent being stuck in local optima), this is a major difficulty in the theoretical analysis of such algorithms. Things seem to become even harder, when (as here) we do not use bit-strings as representations for the individuals. We overcome these problems via what we call *c-trails*. These are hypothetical ways of how to move from one individual to another using simple mutations only. While still some difficulties remain, this allows to analyze the evolutionary algorithms we consider in this chapter. We employ methods similar to the ones we used in Chapter 4 to obtain a tight analysis for the Single Source Shortest Path problem.

We conduct several experiments that show that the proven reduction of the optimization time caused by the use of any of the crossover operators already becomes apparent for small input sizes.

5.2   A Genetic Algorithm for the APSP Problem

Let $G = (V, E)$ be a directed graph with $n := |V|$ vertices and $m := |E|$ edges. Let $w \colon E \to \mathbb{N}$ be a function that assigns to each edge $e \in E$ a weight $w(e)$. Then the APSP problem is to compute a shortest path from every vertex $u \in V$ to every other vertex $v \in V$. A *walk* from u to v is a sequence $u = v_0, v_1, \ldots, v_k = v$ of vertices such that $(v_{i-1}, v_i) \in E$ for all $i \in [1..k]$. The walk is called *path* if it contains each vertex at most once. We will usually describe a walk by the sequence $(e_1, \ldots, e_k)$, $e_i = (v_{i-1}, v_i)$, of edges it traverses. The length of a walk is defined as the sum of the weights of all its edges.

There are two classical algorithms for this problem. The Floyd-Warshall algorithm ([Flo62, War62]) has a cubic runtime and is quite easy to implement. In contrast, Johnson's algorithm [Joh77] is more complicated, but has

5.2 A Genetic Algorithm for the APSP Problem

a superior runtime on sparse graphs. Since the problem is NP-hard [GJ79] if negative cycles exist and simple paths are sought, we will always assume that all weights are non-negative.

One of the strengths of evolutionary computation is that the algorithms are composed of generic components like mutation, crossover and selection. We now give the different components needed for a genetic algorithm that solves the APSP problem.

5.2.1 Individuals and Population

Evolutionary algorithms usually keep a set (population) of solution candidates (individuals), which they gradually improve. In the APSP problem we are aiming for a population containing a shortest path for each pair of distinct vertices. Hence it makes sense to allow paths or walks as individuals. To have more freedom in defining the crossover operator, an individual will simply be a sequence of edges, $(e_1, \ldots, e_k), e_1, \ldots e_k \in E, k \in \mathbb{N}$. However, the selection operator (see below) will ensure that only individuals that are walks can enter the population.

For the APSP problem, a natural choice for the initial population is the set $\mathcal{I} := \{(e) \mid e \in E\}$ of all paths consisting of one edge.

5.2.2 Fitness Function and Selection Operator

The natural choice for the fitness function $f\colon \mathcal{S} \to \mathbb{R}$ is the length of the walk represented by the individual (which in this case has to be minimized). As a result of a crossover operation (see below), we may generate individuals that are not walks. These shall have fitness ∞ and will never be included in the population. Thus, we get as fitness function

$$f(I) = \begin{cases} w(W(u,v)) & \text{if } I \text{ represents a walk from } u \text{ to } v, \\ \infty & \text{otherwise.} \end{cases}$$

with $w(W(u,v))$ being the cost of the walk W from u to v implied by I.

For the APSP problem, diversity is an issue in the sense that we need to end up with one path for each pair of vertices. However, if we ensure directly that the selection operator does not eliminate all paths between a pair of vertices, we can be strict in the selection otherwise. In fact, for each pair (u,v) of vertices our selection operator eliminates all but the fittest individual connecting u to v. Thus, we only need to compare the fitness values of individuals having identical start and end vertices.

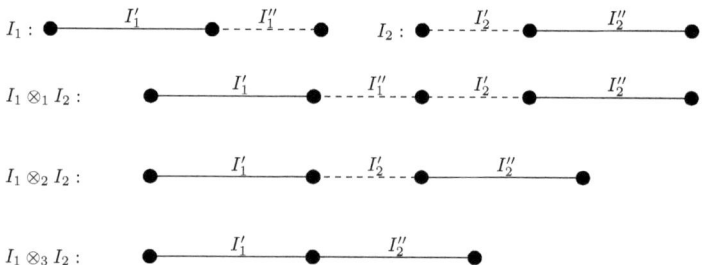

Figure 5.1: The effects of the three crossover operators.

5.2.3 Mutation and Crossover Operators

In evolutionary computation, new individuals are generated by *variation operators*, namely by mutation or crossover (or both).

A mutation operator changes an individual slightly at some random positions. This is done by first choosing a number S at random according to a Poisson distribution $\text{Pois}(\zeta = 1)$ with parameter $\zeta = 1$, making the probability that S is set to k $\Pr[S = k] = \frac{1}{ek!}$. An individual is then mutated by applying the following elementary mutation $S + 1$ times. Let $(u, v) \in E$ be the first edge of the individual and $(u', v') \in E$ be the last edge. Pick an edge e from the set of all edges incident to u or v' uniformly at random. If this edge is (u, v) or (u', v'), remove it from the individual, otherwise append it at the corresponding end of the individual. However, if the individual is a single edge (u, v'), pick an edge uniformly at random from the set of all edges incident to u or v' except (u, v') and append it.

A *crossover* of two individuals combines parts of them to a new individual. In this chapter we consider three variations of the so-called 1-point crossover. For individuals that are bit-strings of length h, it is defined by picking a random position and merging the initial part of the first individual up to the chosen position with the ending part of the second individual starting from the chosen position. Since we do not represent individuals as bit-strings, this cannot be applied directly. Instead, we propose the following three crossover operators to combine two individuals I_1, I_2 containing ℓ_1 and ℓ_2 edges respectively. The crossover operator $\otimes_1$ simply combines both individuals by appending I_2 to I_1. The second operator, $\otimes_2$, chooses a random number $i \in [0..\ell_1]$ and appends I_2 to the first i edges of I_1. Finally, the operator $\otimes_3$ chooses two random numbers $i \in [0..\ell_1]$ and $j \in [0..\ell_2]$. The

5.2 A Genetic Algorithm for the APSP Problem

new individual created by this operator consists of the first i edges of I_1 and the last $\ell_2 - j$ edges of I_2. In Figure 5.1 the effects of the three crossover operators are depicted.

Observe that, unlike mutation, crossover may combine two individuals representing walks to a new individual that no longer represents a walk, and hence has infinite fitness.

5.2.4 $(\mu+1)$-EA$_{\text{apsp}}$ and $(\mu+1)$-GA$_{\text{apsp}}$

The algorithms we consider repeatedly apply variation and selection to a set of individuals. We study both an algorithm that only uses mutation (EA) and an algorithm that uses both mutation and one of the crossover operators (GA).

Both algorithms share the following common framework (cf. also Section 2.5). First, the population $\mathcal{I}$ is initialized. Then, depending on the kind of algorithm, it is decided randomly with a certain probability $p_\otimes$ if a mutation or a crossover step should be done. If a mutation step is done, the algorithm picks an individual uniformly at random from the population and applies the mutation operator to it to generate a new individual. If a crossover step is done, the algorithm picks two individuals uniformly at random from the population and applies a crossover operator to generate a new individual. Afterwards, it checks if there is an individual in the population that connects the same two points as the newly generated individual. If not, the new individual is added to the population. If yes, the old individual is replaced if it is not fitter than the new one. These variation and selection steps are then repeated forever. The pseudocode in Figure 5.2 illustrates this procedure.

If only mutation is used, we get a classical evolutionary algorithm, the so-called $(\mu + 1)$-EA. This means that the population consists of up to μ individuals, and each step one new individual is generated by mutation. Since the population will consist of up to $n(n-1)$ individuals, namely one shortest path candidate for each pair of distinct vertices, $\mu := n(n-1)$ in our case. For sake of simplicity, and to better distinguish this algorithm from the ones analyzed in the other chapters, we will call the algorithm $(\mu + 1)$-EA$_{apsp}$ instead of $(n(n-1)+1)$-EA. If also crossover is used, we get a more general genetic algorithm, which we call $(\mu + 1)$-GA$_{apsp}$.

$(1+1)$-EA for APSP
   Initialization:
1 $\mathcal{I} \leftarrow \{(e) \mid e \in E\}$
2 **repeat**
3   Set $x := 1$ with probability $p_\otimes$
4   **if** $x = 1$
5   **then**
                Crossover:
6   Pick two Individuals $I_1, I_2 \in \mathcal{I}$ u.a.r.
7   Generate a new Individual I' by applying
                      a crossover operator to I_1 and I_2
8   **else**
                Mutation:
9   Pick $I \in \mathcal{I}$ uniformly at random
10   Pick S according to $\Pr[S = k] = \frac{1}{e \cdot k!}$
11   Generate a new Individual I' by $S + 1$
12   times adding or removing an edge from I
        Selection:
13   Let $I'' \in \mathcal{I}$ be the individual with the same
            start-vertex and end-vertex as I', if any.
14   **if** I' is a walk and $w(I') \leq w(I'')$
15   **then**
16   Add I' to $\mathcal{I}$ and remove I'' from $\mathcal{I}$ if it exists.
17 **until** $\mathcal{I}$ is an optimal solution

Figure 5.2: Pseudocode for the two algorithms studied by us. If $p_\otimes < 1$ is a constant greater than zero, both mutation and crossover are used and the resulting algorithm will be called $(\mu + 1)$-GA$_{apsp}$. For $p_\otimes = 0$, only mutation is used as variation operator. We call the resulting algorithm $(\mu + 1)$-EA$_{apsp}$.

5.3 Analysis of the $(\mu+1)$-EA$_{\text{apsp}}$

In this section we show that both in expectation and with high probability[1] the worst case optimization time of the $(\mu + 1)$-EA$_{apsp}$ is $\Theta(n^4)$.

In general, a pair of vertices may be connected by more than one shortest path, and these different paths may consist of different numbers of edges. For our purposes, paths consisting of few edges are more important. To ease the language, we introduce the following notation.

Definition 5.1. *Let $G = (V, E)$ be a graph and let $\ell \in \mathbb{N}$. We define*

$$V_\ell^2 := \{(u, v) \in V^2 \mid u \neq v \text{ and there exists a shortest path} \\ \text{from } u \text{ to } v \text{ consisting of at most } \ell \text{ edges}\}.$$

5.3.1 Upper Bound on the Optimization Time

The main ideas to prove the upper bound of $O(n^4)$ for the $(\mu+1)$-EA$_{apsp}$ are as follows. Being pessimistic, we may assume that shortest paths are found exclusively by adding edges to already found shortest paths, and more specifically, by only adding a single edge in each iteration. Then, to find a shortest path from u to v for $(u,v) \in V_\ell^2$, it suffices that the $(\mu+1)$-EA$_{apsp}$ chooses ℓ times the adequate shortest path already in the solution (with probability $O(n^{-2})$) and adds the appropriate edge (with probability $O(n^{-1})$). If $\ell \geq \log n$, the time needed for this is that sharply concentrated around the mean of $\Theta(\ell n^3)$, that we may use a union bound argument over all $(u,v) \in V_\ell^2$.

Lemma 5.1. *Let $\ell \geq \log(n)$. Within $O(\ell n^3)$ steps, the $(\mu+1)$-EA$_{apsp}$ finds with high probability a shortest path from u to v for all $(u,v) \in V_\ell^2$.*

Proof. Let $(u,v) \in V_\ell^2$. We first analyze the probability that a shortest path from u to v is not found within a certain time. For the analysis, we fix a path $P = ((u, v_1), (v_1, v_2), \ldots, (v_{\ell'-1}, v_{\ell'} = v))$ of length $\ell' \leq \ell$. Note that P will be a technical tool only and we do not aim at finding this particular path.

In the following, we shall only consider mutation steps that perform a single elementary mutation. Note that a mutation consists of a single elementary mutation with probability $\frac{1}{e}$.

---

[1] Recall that "with high probability" means with probability $1 - O(n^{-c})$ for an arbitrary but fixed constant c.

We call a mutation step the *j-th pessimistic improvement* in P if it (i) consists of a single elementary mutation, (ii) creates a shortest path from u to v_{j+1} out of a shortest path from u to v_j that is already in the population, and (iii) the pessimistic improvements $1, \ldots, j-1$ have already been done. Note that this implies that pessimistic improvements appear in ascending order. Obviously, when the $(\mu+1)$-EA$_{apsp}$ has performed the $(\ell'-1)$-st pessimistic improvement in P, a shortest path from u to v has been found.

Let $t := e\eta\ell n^3$ for some $\eta > 2$ and let the random variable t' denote the number of steps the $(\mu+1)$-EA$_{apsp}$ executes until it performs the $(\ell'-1)$-st pessimistic improvement in P. For $i \in [1..t']$ define the random variable X_i by $X_i = 1$ if the i-th mutation step is a pessimistic improvement in P and $X_i = 0$ otherwise. Then, independent of the first $i-1$ steps,

$$\Pr[X_i = 1] \geq \frac{1}{e} \frac{1}{n(n-1)^2} > \frac{1}{en^3} =: p,$$

since (i) the probability that the mutation step consists of a single elementary mutation is $\frac{1}{e}$, (ii) the probability to pick the correct individual is at least $\frac{1}{n(n-1)}$, and (iii) the probability to pick the correct edge is at least $\frac{1}{n-1}$. For $i \in [t'+1..t]$ we independently define X_i by $\Pr[X_i = 1] := p$ and $\Pr[X_i = 0] := 1 - p$. Obviously, X_i is independent of X_j for $1 \leq i < j \leq t$.

Now define the mutually independent binary random variables X_i^* by $\Pr[X_i^* = 1] := p$ and $\Pr[X_i^* = 0] := 1 - p$ for $i \in [1..t]$ and let $X^* := \sum_{i=1}^t X_i^*$. Then the expected value of $X := \sum_{i=1}^t X_i$ is bounded by

$$\mathbb{E}[X] \geq \mathbb{E}[X^*] = pt = \frac{1}{en^3} e\eta\ell n^3 = \eta\ell.$$

If the $(\mu+1)$-EA$_{apsp}$ has not found a shortest path from u to v after t steps, it obviously has not performed the $(\ell'-1)$-st pessimistic improvement in P, and thus $X < \ell' - 1 < \ell'$. Denote by $\mathcal{P}_{uv} := \{P \mid P$ is shortest path from u to $v\}$ the set of shortest paths from u to v. Since for every $i \in [1..t]$ the random variable X_i fulfills $\Pr[X_i = 1 \mid X_1 = x_1, \ldots, X_{i-1} = x_{i-1}] \geq \Pr[X_i = 1]$ for all $x_1, \ldots, x_{i-1} \in \{0, 1\}$, we can use Lemma 3.1 to bound the probability that no shortest path $P \in \mathcal{P}_{uv}$ from u to v is found in t steps by

$$\begin{aligned}
\Pr\left[\text{no } P \in \mathcal{P}_{uv} \text{ found in } t \text{ steps}\right] &\leq \Pr[X < \ell'] \\
&= 1 - \Pr[X \geq \ell'] \\
&\leq 1 - \Pr[X^* \geq \ell'] \\
&= \Pr[X^* < \ell'].
\end{aligned}$$

5.3 Analysis of the $(\mu+1)$-EA$_{apsp}$

Now we can use Theorem 3.1 with $\alpha := \frac{\ell'}{\mathbb{E}[X^*]} \leq \frac{\ell}{\eta\ell} = \frac{1}{\eta}$ to bound this probability of not finding a shortest path from u to v in t steps by

$$\begin{aligned}
\Pr\left[\text{no } P \in \mathcal{P}_{uv} \text{ found in } t \text{ steps}\right] &\leq \Pr[X^* < \ell'] \\
&= \Pr[X^* < \alpha \mathbb{E}[X^*]] \\
&\leq \exp(-\tfrac{1}{2}(1-\alpha)^2 \mathbb{E}[X^*]) \\
&\leq \exp\left(-\tfrac{1}{2}\left(\tfrac{\eta-1}{\eta}\right)^2 \eta\ell\right) \\
&\leq \exp(-\tfrac{1}{8}\eta\ell).
\end{aligned}$$

A simple union bound argument now reveals that the probability that the $(\mu+1)$-EA$_{apsp}$ does not find for all vertex pairs $(u,v) \in V_\ell^2$ a shortest path connecting them ($P \in \mathcal{P}_{uv}$) in t steps is bounded by

$$\begin{aligned}
\Pr\left[\begin{array}{l}\text{there exists } (u,v) \in V_\ell^2 \text{ such that}\\ \text{no } P \in \mathcal{P}_{uv} \text{ is found in } t \text{ steps}\end{array}\right] &\leq \sum_{(u,v) \in V_\ell^2} \Pr\left[\begin{array}{l}\text{no } P \in \mathcal{P}_{uv}\\ \text{found in } t \text{ steps}\end{array}\right] \\
&\leq n(n-1)\exp(-\tfrac{1}{8}\eta\ell) \qquad (5.1)\\
&< n^2 \exp(-\tfrac{1}{8}\eta \log(n)) \\
&= n^{2-\frac{\eta}{8}}.
\end{aligned}$$

For any constant c we can choose $\eta \geq 8(c+2)$. Thus, with probability $1 - O(n^{-c})$ the optimization time is at most $e\eta\ell n^3$. $\square$

Note that we did not try to optimize the constant η. For $\ell = n-1$, Lemma 5.1 yields the following upper bound.

Theorem 5.1. *The optimization time of the $(\mu+1)$-EA$_{apsp}$ is with high probability $O(n^4)$.*

From the strong concentration bound of inequality 5.1 we also derive an $O(n^4)$ bound for the expected optimization time.

Theorem 5.2. *Let $\ell \geq \log(n)$. The expected number of steps until the $(1+1)$-EA finds a shortest path from u to v for all $(u,v) \in V_\ell^2$ is $O(\ell n^3)$. In particular it holds that the expected optimization time of the $(\mu+1)$-EA$_{apsp}$ is $O(n^4)$.*

Proof. Let t_ℓ be the number of steps until the $(1+1)$-EA finds for all $(u,v) \in V_\ell^2$ a shortest path from u to v. In the proof of Lemma 5.1 we showed that the probability that t_ℓ is higher than $e\eta\ell n^3$ is $\Pr[t_\ell > e\eta\ell n^3] \leq n^{2-\frac{\eta}{8}}$. By Lemma 3.2 the expected value $\mathbb{E}[t_\ell]$ of t_ℓ is thus $O(\ell n^3)$. Setting $\ell = n$ we get the upper bound for the expected optimization time. $\square$

Chapter 5: The All-Pairs Shortest Path Problem

Figure 5.3: The complete graph K_n with edge weights w used for the lower bound analysis. The shown edge weights of 1 apply to both directions of the indicated edge. The edges not shown in the figure have weight n.

5.3.2 Lower Bound on the Optimization Time

For the lower bound analysis, we consider the complete directed graph $K_n = ([1..n], \{(u,v) \mid u, v \in [1..n], u \neq v\})$ with edge weights

$$w(u,v) = \begin{cases} 1 & \text{if } |v - u| = 1, \\ n & \text{else.} \end{cases}$$

For two distinct vertices u, v the unique shortest path from u to v is $((u, u+1), \ldots, (v-1, v))$ if $u < v$ and $((u, u-1), \ldots, (v+1, v))$ otherwise (cf. Figure 5.3). These edge weights, together with initialization and selection, ensure that at any time all individuals in the population consist of a single edge or are a shortest path.

Definition 5.2. *The* distance *of two paths is the minimal number of elementary mutations needed to mutate one path into the other. A mutation step* crosses *a distance of c if the path it chooses to mutate and the one it creates by mutation have distance c.*

Note that for the graph K_n with edge weights w the distance of two shortest paths P_1, P_2 is the size of the symmetric difference $\triangle$ of the set of edges $E(P_1), E(P_2)$ of the two paths $|E(P_1) \triangle E(P_2)|$.

Lemma 5.2. *For any $c \in \mathbb{N}$, the probability that an accepted mutation step crosses a distance of c is at most $\frac{4c}{e(n-2)^c} \frac{n-2}{n-3} = O(cn^{-c})$.*

Proof. Let P_1 be the shortest path the mutation step chooses for mutation and P_2 be the shortest path resulting from the mutation step. Each elementary mutation of the mutation step either decreases or increases the distance of the resulting solution to P_2. Hence a shortest path P_2 having a distance of c to P_1 can only be obtained via a sequence of $c + 2i$ elementary mutations for some $i \in \mathbb{N}_0$. In this case, $c + i$ of them decrease and i of them increase the distance of the intermediate solution to P_2. The probability that a certain one of the $c + 2i$ elementary mutations decreases this distance is at

5.3 Analysis of the $(\mu+1)$-EA$_{apsp}$

most $(n-2)^{-1}$, since there are at most 2 additions/deletions that achieve the distance reduction out of at least $2(n-2)$ possible elementary mutations.

Assume in the next two paragraphs that our mutation consists of exactly $c+2i$ elementary mutations. Then there are at most $\binom{c+2i}{i}$ choices for the i elementary mutations that increase the distance to P_2. In consequence, the probability to end up with P_2 is at most $\binom{c+2i}{i}(n-2)^{-(c+i)}$.

It is easy to see that there are at most $2c$ shortest paths P_2 that have exactly a distance of c to P_1. Thus, if the mutation step performs $c+2i$ elementary mutations, the probability to end up with any shortest path P_2 having a distance of c to P_1 is at most $2c\binom{c+2i}{i}(n-2)^{-(c+i)}$.

Recall that the probability that our mutation consists of $c+2i$ elementary mutations is $(e(c+2i-1)!)^{-1}$. Hence the probability that a single mutation step crosses a distance of c is at most

$$\sum_{i=0}^{\infty} \frac{1}{e(c+2i-1)!} \binom{c+2i}{i} \frac{2c}{(n-2)^{c+i}} = \frac{2c}{e(n-2)^c} \sum_{i=0}^{\infty} \frac{c+2i}{i!(c+i)!} \frac{1}{(n-2)^i}$$

$$\leq \frac{4c}{e(n-2)^c} \sum_{i=0}^{\infty} \frac{1}{(n-2)^i}$$

$$\leq \frac{4c}{e(n-2)^c} \frac{n-2}{n-3}$$

$$= O(cn^{-c}).$$

□

Lemma 5.3. *For any constant c_1, there exists a constant $c := c_1+3$ such that with probability $1 - O(n^{-c_1})$, during its optimization time the $(\mu+1)$-EA$_{apsp}$ will only accept mutation steps that cross a distance of at most c.*

Proof. We know from Theorem 5.1 that for any arbitrary but fixed constant c_1 the $(\mu+1)$-EA$_{apsp}$ has with probability $1-O(n^{-c_1})$ an optimization time of $O(n^4)$ and from Lemma 5.2 that a mutation step that crosses a distance of c_2 is accepted with probability $O(c_2 n^{-c_2})$. Thus, the probability that during the optimization time of the $(\mu+1)$-EA$_{apsp}$ a mutation step crossing a distance of exactly c_2 is accepted is at most

$$(1 - O(n^{-c_1})) \cdot O(n^4) \cdot O(c_2 n^{-c_2}) + O(n^{-c_1}) = O(c_2 n^{4-c_2}) + O(n^{-c_1}).$$

Equivalently, the probability that during the optimization time of the $(\mu+1)$-EA$_{apsp}$ an accepted mutation step crosses a distance of more than c

is bounded by

$$O(n^{-c_1}) + \sum_{c_2=c+1}^{\infty} O(c_2 n^{4-c_2}) = O(n^{-c_1}) + \sum_{c_2=c_1}^{\infty} O(c_2 n^{-c_2}) = O(n^{-c_1}).$$

□

Let $P^* := ((1,2),(2,3),\ldots,(n-1,n))$ be the shortest path from 1 to n in K_n with edge weights w. Consider a sequence of mutation steps (each changing at least one edge) that may create P^*. Of these steps consider the last $\lfloor \frac{n-3}{c} \rfloor$ where c is the constant from Lemma 5.3. Let the paths that are created during these steps be $P_0, P_1, \ldots P_{\lfloor \frac{n-3}{c} \rfloor} = P^*$. Since P^* consists of $|P^*| = n-1$ edges and since P_j has with high probability at most c edges more than P_{j-1}, we have that $|P_j| \geq 2$ for all $j \in [0..\lfloor \frac{n-3}{c} \rfloor]$ and thus all P_j are shortest paths. Thus, these paths fulfill the requirements of the following definition.

Definition 5.3 (c-Trail). *A c-trail $T := (P_0, P_1, \ldots, P_{\lfloor \frac{n-3}{c} \rfloor})$ of P^* is a sequence of shortest paths such that P_0 consists of at least 2 edges, $P_{\lfloor \frac{n-3}{c} \rfloor} = P^*$, and for all $j \in [1..\lfloor \frac{n-3}{c} \rfloor]$ P_{j-1} and P_j have a distance of at most c.*

Since there are at most $(2c)^2$ shortest paths that have a positive distance of at most c from P_j, there are at most $(4c^2)^{\lfloor \frac{n-3}{c} \rfloor}$ such c-trails.

Theorem 5.3. *The optimization time of the $(\mu+1)$-EA$_{apsp}$ on K_n with edge weights w is with high probability $\Omega(n^4)$.*

Proof. Let c be the constant from Lemma 5.3. In order to create P^* the $(\mu+1)$-EA$_{apsp}$ has to perform all $\lfloor \frac{n-3}{c} \rfloor$ mutation steps that create P_j out of P_{j-1} for $j \in [1..\lfloor \frac{n-3}{c} \rfloor]$ of one of the c-trails of P^* (and the mutation steps leading to P_0, which we will ignore in this proof). First, we will analyze the number of steps the $(\mu+1)$-EA$_{apsp}$ needs to follow one particular c-trail of P^*. Then, we will prove that with high probability the $(\mu+1)$-EA$_{apsp}$ will not follow any of the c-trails of P^* in less than $\Omega(n^4)$ steps.

Fix one c-trail $T = (P_0, P_1, \ldots, P_{\lfloor \frac{n-3}{c} \rfloor})$ of P^*. We call a mutation step an *improvement* in T if it creates P_j out of P_{j-1} for some $1 \leq j \leq \lfloor \frac{n-3}{c} \rfloor$. If all $\lfloor \frac{n-3}{c} \rfloor$ improvements in T have been done, we say that the $(\mu+1)$-EA$_{apsp}$ has followed T.

Let $t := \frac{1}{80c^4}(n-1)^4$ and let the random variable t' denote the number of steps the $(\mu+1)$-EA$_{apsp}$ needs to follow T. For $i \in [1..t']$ define the binary

5.3 Analysis of the $(\mu+1)$-EA$_{apsp}$

random variables X_i by $X_i = 1$ if the i-th mutation step is an improvement in T. An improvement changes at least 1 and at most c edges of a path. In order to change c' edges, it first has to pick the right individual with probability $\frac{1}{n(n-1)}$ and then change the c' edges with probability $\frac{4c'}{e(n-2)^{c'}}\frac{n-2}{n-3}$ (cf. Lemma 5.2). Thus, for $n \geq 6$, the probability that $X_i = 1$ is independent of the steps before bounded by

$$\Pr[X_i = 1] \leq \sum_{c'=1}^{c} \frac{1}{n(n-1)} \frac{4c'}{e(n-2)^{c'}} \frac{n-2}{n-3}$$

$$\leq \frac{4}{en(n-1)(n-2)} \cdot \frac{n-2}{n-3} \cdot \sum_{c'=0}^{c-1} \frac{c'}{(n-2)^{c'}}$$

$$< \frac{4c}{en(n-1)(n-2)} \cdot \frac{n-2}{n-3} \cdot \frac{n-2}{n-3}$$

$$< \frac{8c}{e(n-1)^3}.$$

Let $p := \frac{8c}{e(n-1)^3}$. For $t' < i \leq t$ define the binary random variable X_i by $\Pr[X_i = 1] := p$ and $\Pr[X_i = 0] := 1 - p$. As needed in Lemma 3.1, for all $i \in [1..t]$ and all $x_1, \ldots, x_{i-1} \in \{0,1\}$ it holds that $\Pr[X_i = 1 \mid X_1 = x_1, \ldots, X_{i-1} = x_{i-1}] \leq p$. Thus, we define the binary random variables X_i^* by $\Pr[X_i^* = 1] := p$ and $\Pr[X_i^* = 0] := 1 - p$.

The expected value of $X^* := \sum_{i=1}^{t} X_i^*$ is

$$\mathbb{E}[X^*] = \sum_{i=1}^{t} \Pr[X_i^* = 1] = pt = \frac{8c}{e(n-1)^3} \frac{1}{80c^4}(n-1)^4 = \frac{n-1}{10ec^3}.$$

If the $(\mu+1)$-EA$_{apsp}$ has found P^* in t steps by following the c-trail T, then obviously $X := \sum_{i=1}^{t} X_i \geq |T| = \lfloor \frac{n-3}{c} \rfloor$. Hence,

$$\Pr[P^* \text{ found in } t \text{ steps by following } T] = \Pr[X \geq |T|].$$

Let $\beta := \frac{|T|}{\mathbb{E}[X^*]}$. Then for $n \geq 5 + 2c$ it holds that

$$\beta = \lfloor \frac{n-3}{c} \rfloor \cdot \frac{10ec^3}{n-1} \geq \frac{n-3-c}{c} \cdot \frac{2c}{n-1} \cdot 5ec^2 \geq 5ec^2.$$

Hence, by Lemma 3.1 and Theorem 3.1, the probability of finding P^* in

$t = \frac{1}{80c^4}(n-1)^4$ steps by following c-trail T is bounded by

$$\begin{aligned}
\Pr[X \geq |T|] &\leq \Pr[X^* \geq |T|] \\
&= \Pr[X^* \geq \beta \mathbb{E}[X^*]] \\
&< (e^{\beta-1}\beta^{-\beta})^{\mathbb{E}[X^*]} \\
&\leq \left(\frac{e}{\beta}\right)^{\beta \mathbb{E}[X^*]} \\
&\leq (5c^2)^{-|T|} \\
&= (5c^2)^{-\lfloor \frac{n-3}{s} \rfloor}.
\end{aligned}$$

Since the $(\mu+1)$-EA$_{apsp}$ has to follow one of the c-trails of P^* in order to find P^*, the probability that the $(\mu+1)$-EA$_{apsp}$ finds P^* in $t = \frac{1}{80c^4}(n-1)^4$ steps is bounded by

$$\begin{aligned}
\Pr[P^* \text{ found in } t \text{ steps}] &\leq \sum_{T \in \mathcal{T}} \Pr[P^* \text{ found in } t \text{ steps by following } T] \\
&\leq \sum_{T \in \mathcal{T}} (5c^2)^{-\lfloor \frac{n-3}{c} \rfloor} \\
&= \left(\frac{4}{5}\right)^{\lfloor \frac{n-3}{c} \rfloor}.
\end{aligned}$$

Here $\mathcal{T}$ denotes the set of all c-trails of P^*. In the penultimate line we used the fact that there are at most $(4c^2)^{\lfloor \frac{n-3}{c} \rfloor}$ c-trails of P^*. Since the $(\mu+1)$-EA$_{apsp}$ has to find P^* to solve the APSP it needs with high probability at least $\Omega(n^4)$ steps. □

Observe that this theorem implies an expected optimization time of $\Omega(n^4)$.

5.4 Upper Bound on the Optimization Time of the $(\mu+1)$-GA$_{apsp}$

We now prove that if we use the $(\mu+1)$-GA$_{apsp}$ for the APSP problem, that is, we enrich the $(\mu+1)$-EA$_{apsp}$ with a crossover operator, then the expected optimization time drops to $O(n^{3.5+\varepsilon})$ for any $\varepsilon > 0$. This bound holds for any constant crossover probability $0 < p_\otimes < 1$.

While it seems natural that the additional use of powerful variation operators should speed up computation, this behavior could so far not be proven

5.4 Upper Bound on the Optimization Time of the $(\mu+1)$-GA_{apsp}

for a non-artificial problem. Several reasons for this have been discussed in the literature. In our setting, the following aspect seems crucial. The hoped for strength of the crossover operator lies in the fact that it can advance a solution significantly. E. g., it can combine two shortest paths consisting of ℓ_1 and ℓ_2 edges to one consisting of $\ell_1 + \ell_2$ edges in one operation. On the negative side, for this to work, the two individuals we try to combine have to fit together. Thus with relatively high probability, the crossover operator will produce an invalid solution (here, no path at all). Often, this disadvantage seems to outnumber the chance of faster progress.

Our analysis shows that this does not happen in our setting. In fact, from the point on when our population contains all shortest paths having $O(n^{1/2+\varepsilon})$ edges, crossover becomes so powerful that we would not even need mutation anymore.

We can prove the claimed upper bound for all three crossover operators introduced in Section 5.2.3. However, as the crossover operators we use become more elaborate, for the proof we need to comply to the following restrictions.

R1: Among two shortest paths the fitness function prefers the one consisting of fewer edges. (Needed for $\otimes_2$.)

R2: The input graph has unique shortest paths. (Needed for $\otimes_3$.)

Given these restrictions, we show for each crossover operator a certain probability that it successfully creates a longer path by combining two shorter paths. Using these success probabilities we prove the expected optimization time of $O(n^{3.5+\varepsilon})$.

Lemma 5.4. *Let $k > 1$. Assume the population $\mathcal{I}$ contains a shortest path for any pair of vertices $(u', v') \in V_k^2$. Let $\ell \in [k+1..2k]$ and $(u,v) \in V_\ell^2$. Then the following holds.*

a) *A single execution of the $\otimes_1$-operator generates a shortest path from u to v with probability $\Omega(\frac{2k+1-\ell}{n^4})$.*

b) *Assume R1. A single execution of the $\otimes_2$-operator generates a shortest path from u to v having at most ℓ edges with probability $\Omega(\frac{(2k+1-\ell)^2}{kn^4})$.*

c) *Assume R2. A single execution of the $\otimes_3$-operator generates the shortest path from u to v with probability $\Omega(\frac{(2k+1-\ell)^3}{k^2 n^4})$.*

Chapter 5: The All-Pairs Shortest Path Problem

Proof. a) The $\otimes_1$-operator can generate a shortest path from u to v by picking a path P_u starting in u and a path P_v ending in v, such that P_u together with P_v forms a path from u to v. A particular pair (P_u, P_v) is chosen with probability at least

$$(n(n-1))^{-2} = \Omega(\tfrac{1}{n^4}).$$

This leaves the task of counting the number of pairs that generate a shortest path from u to v. Let $P = ((u, w_1), (w_1, w_2), \ldots, (w_{\ell-1}, v))$ be a shortest path from u to v having ℓ edges. Then, for every vertex $w_i, i \in [\ell - k, k]$, a shortest path from u to w_i and a shortest path from w_i to v are in the population. Hence, there are at least $2k + 1 - \ell$ pairs of paths that the $\otimes_1$-operator can combine to a shortest path from u to v. In summary, the probability that a single crossover step generates a shortest path from u to v is at least $\Omega(\tfrac{2k+1-\ell}{n^4})$.

b) To generate a shortest path from u to v, it suffices that the $\otimes_2$-operator picks a path P_u starting in u, a path P_v ending in v, and a number $i \in [0..|P_u|]$ such that the first i edges of P_u together with P_v form a path from u to v. The probability that a particular triple (P_u, P_v, i) with $|P_u| \leq k, |P_v| \leq k, i \leq |P_u|$ is chosen is at least

$$(n(n-1))^{-2}(k+1)^{-1} = \Omega(\tfrac{1}{kn^4}).$$

It remains to count how many such triples generate a shortest path from u to v. Let $P = ((u, w_1), \ldots, (w_{\ell-1}, v))$ be such a shortest path having ℓ edges. Let $\ell - k \leq j \leq k$. Then $\mathcal{I}$ contains a shortest path $P_u = ((u, w'_1), \ldots, (w'_{j-1}, w_j))$ from u to w_j having j edges. For each $i \in [\ell - k..j]$, $\mathcal{I}$ contains a shortest path P_v from w'_i to v, since $\ell - i \leq k$. Obviously, the first i edges of P_u combined with P_v form a shortest path from u to v. Hence, the total number of triples yielding a shortest path from u to v having ℓ edges is at least

$$\sum_{j=\ell-k}^{k} (j - (\ell - k) + 1) = \Omega((2k + 1 - \ell)^2).$$

Thus, the probability that $\otimes_2$ generates such a path in a single step is at least $\Omega(\tfrac{(2k+1-\ell)^2}{kn^4})$.

c) To generate P, the $\otimes_3$-operator has to pick a path P_u starting in u, a path P_v ending in v, and numbers $i \in [0..|P_u|], j \in [0..|P_v|]$ such that the first i edges of P_u together with the last j edges of P_v form the path P. The probability that a particular 4-tuple (P_u, P_v, i, j) with $|P_u| \leq k, |P_v| \leq k, i \leq |P_u|, j \leq |P_v|$ is chosen is at least

$$(n(n-1))^{-2}(k+1)^{-2} = \Omega(\tfrac{1}{k^2 n^4}).$$

5.4 Upper Bound on the Optimization Time of the $(\mu+1)$-GA_{apsp}

It remains to count the number of such 4-tuples that generate P. For this, consider two sub-paths of P, one starting at u, the other ending at v. Observe that those sub-paths are also shortest paths. Since we assume all shortest paths to be unique, both sub-paths will be in the population if they consist of at most k edges. If the sum of the numbers of edges of both paths is some $i \in [\ell..2k]$, they have $i - \ell$ edges in common and the number of successful crossover positions is $i - \ell + 1$. The number of pairs of sub-paths that have $i - \ell$ edges in common is $2k + 1 - i$. Hence, the total number of 4-tuples yielding P is at least

$$\sum_{i=\ell}^{2k}(i - \ell + 1) \cdot (2k + 1 - i) = \sum_{i=0}^{2k-\ell}(i+1) \cdot (2k+1-i-\ell)$$
$$= \Omega((2k+1-\ell)^3).$$

Thus, the probability that $\otimes_3$ generates the shortest path P in a single step is at least $\Omega(\frac{(2k+1-\ell)^3}{k^2 n^4})$. □

Corollary 5.1. *Let $k > 1$ and $\ell = \frac{3k}{2}$. Assume the population $\mathcal{I}$ contains for any pair of vertices $(u', v') \in V_k^2$ a shortest path. Assuming R1 for $\otimes_2$ and R2 for $\otimes_3$ the following holds.*

a) *Let $(u, v) \in V_\ell^2$. A single execution of the $\otimes_i$-operator for $i \in \{1, 2, 3\}$ will create a shortest path from u to v with probability at least $\Omega(\frac{k}{n^4})$.*

b) *The expected number of crossover steps until $\mathcal{I}$ contains for all $(u, v) \in V_\ell^2$ a shortest path from u to v is $O(\frac{n^4 \log(n)}{k})$.*

Proof. a) This follows directly by plugging ℓ into Lemma 5.4.

b) This proof is similar to the proof of the coupon collector's theorem (cf. [AS00]). Let $r = |V_\ell^2| - |V_k^2| = O(n^2)$ be the number of paths that have to be found. By a) the first of the sought after paths will be found after an expected number of $O(\frac{n^4}{k} \frac{1}{r})$ steps. If i paths have been found, it will take an expected number of $O(\frac{n^4}{k} \frac{1}{r-i})$ steps until the $(i+1)$-st path is found. Hence, to find all r paths takes

$$\sum_{i=0}^{r-1} O\left(\frac{n^4}{k}\right) \frac{1}{r-i} = O\left(\frac{n^4}{k}\right) \sum_{i=1}^{r} \frac{1}{i} = O\left(\frac{n^4 \log(n)}{k}\right)$$

steps. □

Theorem 5.4. *Let $i \in [1..3]$. If the conditions for the $\otimes_i$-operator hold, then the $(\mu + 1)$-GA_{apsp} using mutation and $\otimes_i$-crossover with any constant crossover probability $0 < p_\otimes < 1$ needs an expected number of $O(n^{3.5}\sqrt{\log(n)})$ steps to solve the APSP problem.*

Proof. Let $k := \sqrt{n \log(n)}$. Both the $\otimes_i$ and the mutation operator happen with constant probability and neither can decrease the fitness of the population. Thus, for an upper bound we may consider the steps of one of the operators only. Considering the steps of the mutation operator only, according to Theorem 5.2, the algorithm will need in expectation at most $O(n^{3.5}\sqrt{\log(n)})$ steps to find for every $(u,v) \in V_k^2$ a shortest path from u to v. (Note that Theorem 5.2 also holds if a fitness function preferring fewer edges is used.) To find the remaining shortest paths, we only consider the steps of the $\otimes_i$-operator and apply Corollary 5.1 repeatedly until $\ell = n - 1$. Hence the expected number of steps is

$$\sum_{j=\lfloor \log_c(k) \rfloor}^{\lceil \log_c(n) \rceil} O\left(\frac{n^4 \log(n)}{c^j}\right) = O\left(n^4 \log(n) \sum_{j=\lfloor \log_c(k) \rfloor}^{\lceil \log_c(n) \rceil} \frac{1}{c^j}\right)$$

$$= O\left(\frac{n^4 \log(n)}{c^{\log_c(k)}} \sum_{j=0}^{\lceil \log_c(\frac{n}{k}) \rceil} \frac{1}{c^j}\right)$$

$$= O\left(n^{3.5}\sqrt{\log(n)}\right)$$

where $c := \frac{3}{2}$. □

Necessity of the Restrictions

We now demonstrate where the proof of the optimization time would fail without the additional constraints for $\otimes_2$ and $\otimes_3$.

To see the necessity of assumption R1 (the fitness function prefers paths consisting of fewer edges), consider for even n the complete graph $K_n = ([1..n], \{(u,v) \mid u,v \in [1..n], u \neq v\})$ with edge weights

$$w'(u,v) = \begin{cases} 1 & \text{if } |v-u| = 1 \text{ and } u,v \leq \frac{n}{2} + 1\}, \\ \frac{2}{n} & \text{if } |v-u| = 1 \text{ and } u,v \geq \frac{n}{2} + 2\} \\ \frac{2}{n} & \text{if } (u,v) \in \{(2, \frac{n}{2}+2), (\frac{n}{2}+2, 2), (n,1), (1,n)\} \\ 1 + w_{uv}^2 & \text{else} \end{cases}$$

5.4 Upper Bound on the Optimization Time of the $(\mu+1)$-GA$_{apsp}$

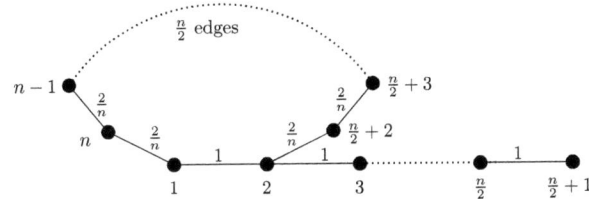

Figure 5.4: The complete graph K_n with edge weights w' for which the analysis of the $\otimes_2$-operator fails if the fitness function does not prefer individuals with fewer edges. The shown edge weights apply to both directions of the indicated edge. The edges not shown in the figure are longer than the shortest paths shown.

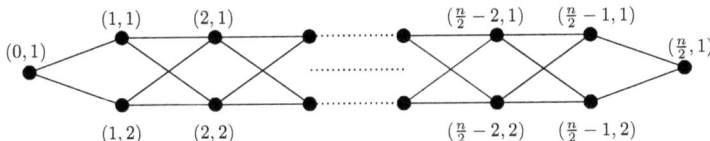

Figure 5.5: The complete graph K_n'' with edge weights w'' for which the analysis of the $\otimes_3$-operator fails, since the shortest paths are not unique. The edges shown in the figure have weight 1 in both directions and the ones not depicted are longer than the shortest paths shown.

depicted in Figure 5.4. Here, w_{uv} is the cost of the shortest path using the edges of weight 1 and $\frac{2}{n}$ from u to v.

Assume, as in Lemma 5.4, that for all vertex pairs $(u,v) \in V_k^2$ a shortest path is in the population $\mathcal{I}$, and that $\ell \in [k+1..2k]$ and $\ell \leq \frac{n}{2}$. Now consider the computation of a shortest path from $u := 1$ to $v := \ell + 1$ using the $\otimes_2$-operator. Two such shortest paths exist, namely P_1 which uses the edge $(1,2)$ of cost 1 and has ℓ edges and P_2 which uses the $\frac{n}{2}$ edges of cost $\frac{2}{n}$ and has $\ell - 1 + \frac{n}{2}$ edges. If $\mathcal{I}$ contains for the paths from u to i for $i \in [2..k+1]$ the paths using the edge $(1,2)$, the proof of Lemma 5.4b) works. However, if $\mathcal{I}$ contains the paths using the $\frac{n}{2}$ edges of cost $\frac{2}{n}$, the probability that the $\otimes_2$-operator picks a convenient triple (P_u, P_v, i) drops from $\Omega(\frac{1}{kn^4})$ to $\Omega(\frac{1}{n^5})$ since there are $\Omega(n)$ possible positions to cut P_u.

The assumption R2 that the shortest paths are unique is essential for the proof for the $\otimes_3$-operator. To see this, consider for even n the complete graph $K_n'' := (V, \{(u,v) \mid u,v \in V, u \neq v\})$ with $V := [1..\frac{n}{2} - 1] \times \{1,2\} \cup$

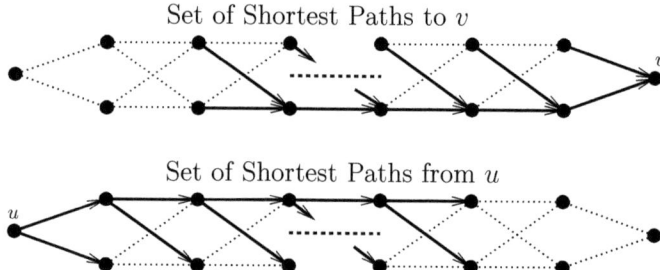

Figure 5.6: An example for sets of shortest paths in K_n'' that do not overlap enough and thus do not fulfill the requirements for the proof of Lemma 5.4c).

$\{(0,1), (\frac{n}{2}, 1)\}$ and with edge weights

$$w''(u,v) = \begin{cases} 1 & \text{if } |v_1 - u_1| = 1, \\ 1 + w_{uv}^2 & \text{else} \end{cases}$$

depicted in Figure 5.5. Similar to above, w_{uv} is the length of the shortest path using the edges of weight 1 from u to v. Observe that there are many different shortest paths connecting two vertices, all having an equal number of edges.

Assume that $\mathcal{I}$ contains the shortest paths from $u := (0, 1)$ to i for $i \in [1..k] \times \{0, 1\}$ and from j to $v := (\frac{n}{2}, 1)$ for $j \in [\frac{n}{2} - k..\frac{n}{2}] \times \{0, 1\}$ as given in Figure 5.6. Then for any shortest path from u to v (having $\ell := \frac{n}{2}$ edges) the population will not contain all sub-paths of length up to k, as needed by Lemma 5.4. Even more, any pair of paths, one starting in u, the other ending in v, will only overlap on at most two vertices.

5.5 Experimental Results

In the previous sections we saw that the asymptotic worst case optimization time of the $(\mu + 1)$-EA$_{apsp}$ is $\Theta(n^4)$, while that of the $(\mu + 1)$-GA$_{apsp}$ is $O(n^{3.5+\varepsilon})$. To show that this difference is in fact noticeable in practice, we implemented the algorithm given in Section 5.2.4 with the three different crossover operators and ran it on the following three graph classes.

The first graph class are the weighted complete graphs K_n with edge weights w from Section 5.3.2, that have edge weights 1 for all edges (u, v) with

5.5 Experimental Results

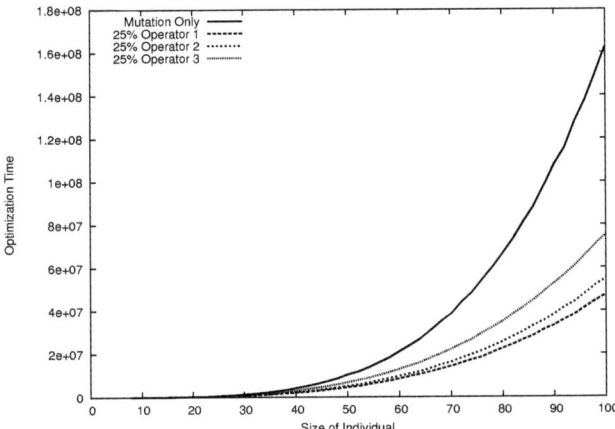

Figure 5.7: Optimization time for the various crossover operators on the complete graph K_n with edge weights w (see Section 5.3.2).

$|v-u| = 1$, and weight n for all other edges. The second and third graph class are the complete graphs K_n with edge weights w' and the complete graphs K_n'' with edge weights w'' used in Section 5.4 to show why we need additional assumptions in the proofs concerning the operators $\otimes_2$ and $\otimes_3$. All edge weights of w' have been multiplied by $\frac{n}{2}$ to ensure their integrality. Note that, although we put restrictions on $\otimes_2$ and $\otimes_3$ in the proofs, our implementation does not prefer paths with fewer edges nor does it need unique shortest paths when applying $\otimes_3$.

We ran the implementation of our algorithm on all three graph classes mentioned above, once using mutation only and once for each crossover operator with crossover probability $\frac{1}{4}$. For all graph classes we considered the graphs having an even number of vertices between 8 and 100. On each instance the algorithm was run 50 times. The average optimization times for the experiments are shown in Figure 5.7, Figure 5.8, and Figure 5.9. To keep the plots legible we did not plot the standard deviations. However, they are below 10% for all instances of 40 or more edges. It can clearly be seen that adding any of the crossover operators does speed up the computation considerably. The results also show that the "bad graphs" K_n with edge weights w' and K_n'' with edge weights w'' from Section 5.4 are not hard to

Chapter 5: The All-Pairs Shortest Path Problem

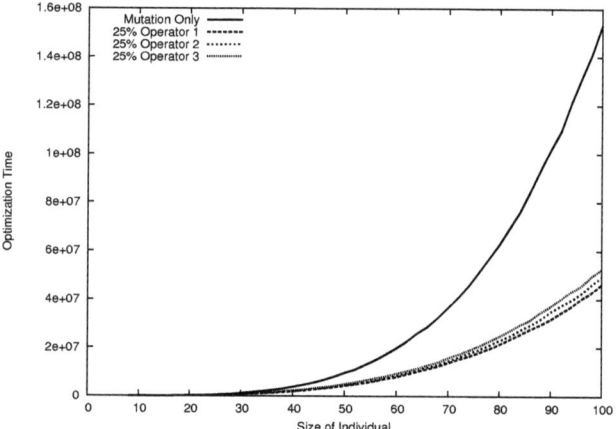

Figure 5.8: Optimization time for the various crossover operators on the complete graph K_n with edge weights w' (see Figure 5.4).

solve for the corresponding crossover operators. In comparison to the other graph classes, the mutation operator is more effective on K_n'' with w''. The reason is probably that due to the structure of w'' the mutation operator has a lot of possibilities to create shortest paths. Thus, the difference between runs with and without crossover are not quite so noticeable.

To estimate the different exponents of the runtimes with and without crossover, we additionally ran the algorithm 20 times each on instances of size $50, 60, 70, \ldots, 250$. We chose these bigger input sizes to weaken the effect of the lower order terms of the runtime. To see the different exponents, we use log-log plots[2]. For any polynomial $f(x) = ax^n + \mathrm{o}(x^n)$, a log-log plot will plot the function

$$\log\left(f(\log^{-1}(x))\right) = \log\left(a(e^x)^n + \mathrm{o}((e^x)^n)\right) = nx + \mathrm{o}(x)$$

hence exposing the exponent of $f(x)$. Figure 5.10, Figure 5.11, and Figure 5.12 show the log-log plots. The difference in the exponent of the runtime between the mutation-only algorithm and any of the algorithms using crossover can easily be discerned in the plots. We also calculated the slope of

---

[2] In other words, both the x and the y-axis are scaled logarithmically.

5.5 Experimental Results

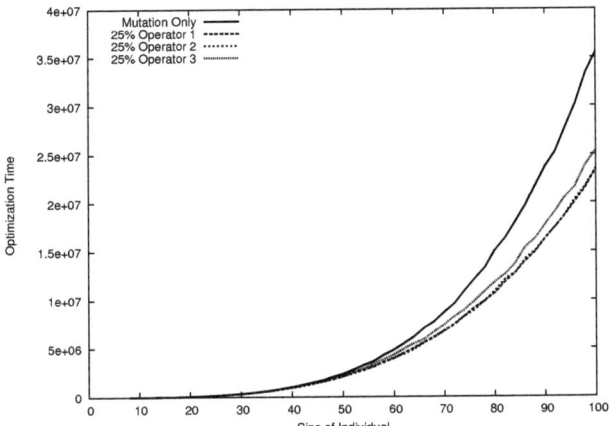

Figure 5.9: Optimization time for the various crossover operators on the complete graph K_n'' with edge weights w'' (see Figure 5.5).

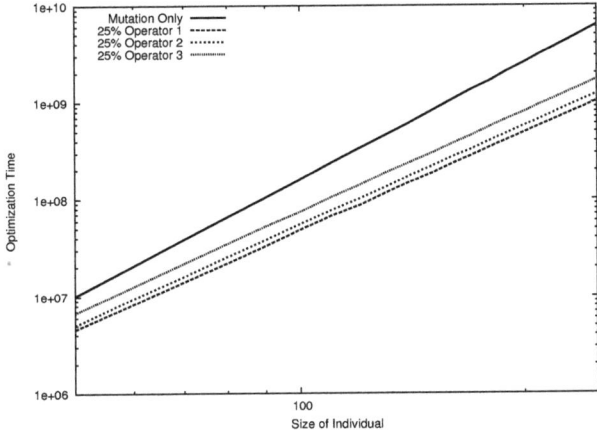

Figure 5.10: Log-log plots for K_n with edge weights w.

Chapter 5: The All-Pairs Shortest Path Problem

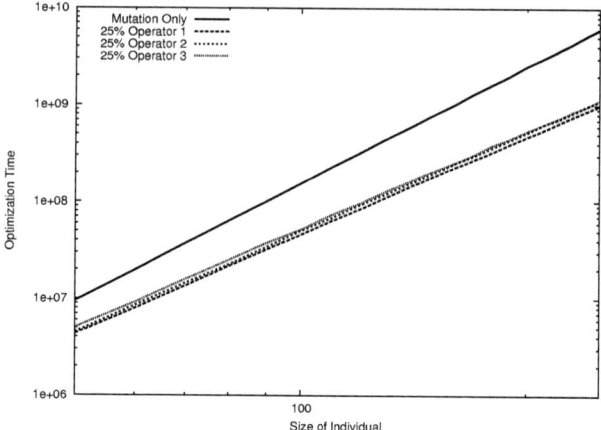

Figure 5.11: Log-log plots for K_n with edge weights w'.

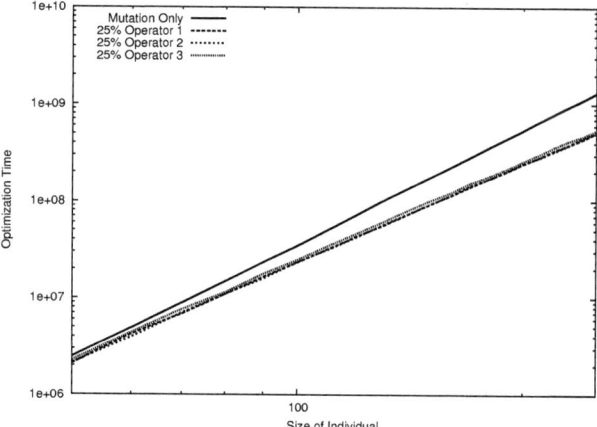

Figure 5.12: Log-log plots for K_n'' with edge weights w''.

the plots. Figure 5.13 shows the results of these calculations. The numbers for K_n using w and w' and K_n'' using w'' show that when using only mutation some graphs may indeed cause a quartic runtime. Also, on all three examples crossover seems to be slightly faster than the $O(n^{3.5})$ suggested by our upper bound.

	w	w'	w''
Mutation Only	4.00	4.01	3.90
Crossover ($\otimes_1$)	3.37	3.38	3.43
Crossover ($\otimes_2$)	3.41	3.42	3.41
Crossover ($\otimes_3$)	3.44	3.36	3.41

Figure 5.13: The slope of the log-log plots in Figure 5.10, Figure 5.11 and Figure 5.12.

The experiments also show that $\otimes_1$ seems to have a slight edge over $\otimes_2$ which in turn is slightly faster than $\otimes_3$. We conjecture that this is caused by the fact that the simpler crossover operators on average combine longer paths than the more complicated ones.

5.6 Summary

In this chapter, we presented the first non-artificial problem for which a natural evolutionary algorithm using only mutation is provably outperformed by one using mutation and crossover. By a rigorous analysis of the optimization time, we proved that the All-Pairs Shortest Path problem can be solved by an evolutionary algorithm using crossover in an expected optimization time of $O(n^{3.5+\varepsilon})$, whereas the corresponding algorithm using only mutation needs an expected optimization time of $\Omega(n^4)$ in the worst case. While this clearly does not beat the best classical algorithm custom tailored for the All-Pairs Shortest Path problem, this result does give a better theoretical foundation for the use of crossover in practical applications than previous results on artificially defined pseudo-boolean functions.

Chapter 6

Sorting

This Chapter is based on the paper "Directed Trees: A Powerful Representation for Sorting and Ordering Problems" by Benjamin Doerr and E. H [DH08].

6.1 Introduction

Evolutionary algorithms, as all randomized search heuristics, are composed of generic, reusable parts (e.g., representations, mutation operators, fitness functions). An expert in such methods can hopefully solve algorithmic problems easily by plugging together appropriate generic components without fully analyzing the problem itself. To this aim suitable representations and mutation operators must be known. A series of papers on the Euler tour problem [DHN06, DKS07, DJ07] demonstrates how more adequate representations yield better algorithms.

In this chapter, we develop a new representation for permutations which is based on trees. This representation admits a natural mutation operator. Building on this framework, we obtain a natural $(1 + 1)$ evolutionary algorithm for the classical problem of sorting n elements. As we shall see, this algorithm is significantly faster than previous evolutionary approaches to the Sorting problem. Additionally, by distinguishing between wrong and unknown information, we may extract some reliable information already before the algorithm has found the terminal solution.

6.1.1 Related Work

To the best of our knowledge, there is only one theoretical investigation on how to solve the Sorting problem via evolutionary means. Scharnow, Tinnefeld, and Wegener [STW04] designed a $(1+1)$ evolutionary algorithm for the Sorting problem based on the following components.

As search space they use the union of all possible permutations $\pi = (\pi(1), \ldots, \pi(n))$ of the elements. For this representation, they propose the following two mutation operators.

- EXCHANGE(i, j) swaps the elements in positions i and j.

- JUMP(i, j) places the element in position i in position j and moves the elements in between one position towards i.

To determine how close the current solution is to being sorted, they use the following well-known measures of presortedness known in adaptive Sorting [PM95] as fitness functions.

- HAM(π) is the number of elements in the correct position (the Hamming distance).

- EXC(π) is the number of exchanges necessary to sort the sequence.

- INV(π) is the number of pairs of elements $(\pi(i), \pi(j))$ that are in the wrong order.

- LAS(π) is the length of the longest ascending subsequence.

- RUN(π) is the number of maximal sorted blocks (called runs).

For this $(1+1)$ evolutionary algorithm using any fitness function expect RUN they give a lower bound on the expected optimization time of $\Omega(n^2)$ independent of whether EXCHANGE, JUMP, or both mutation operators are used. They prove an expected upper bound of $O(n^2 \log n)$ for all fitness functions except RUN, which holds when both mutation operators are used. However, for most fitness functions only one operator is used in the proof. For the combinations of the fitness functions HAM and EXC with the mutation operator EXCHANGE and of the fitness function INV with either EXCHANGE or JUMP they give a tight bound of $\Theta(n^2 \log n)$. For the fitness function RUN they propose an expected exponential optimization time if only JUMPs are used.

6.1 Introduction

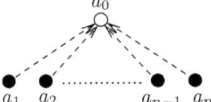

Figure 6.1: The initial solution. The elements of $\mathcal{G} = \{a_1, \ldots, a_n\}$ are incomparable in this order.

6.1.2 Our Results

While a sorted sequence of elements can conveniently be represented by a permutation, intermediate results of many Sorting algorithms cannot. Therefore, a more natural view of Sorting is that we start with an unsorted set of elements and successively by comparing elements add order to the set. Hence the approach we propose in this chapter is to use a sufficiently rich set of orders on the ground set of elements to be sorted as search space. This should include the empty order as natural initial solution and all linear orders (permutations) as possible final solutions. The advantage of a search space built on this paradigm is that by punishing incorrectly ordered element pairs in the fitness function, we can easily and in a natural manner ensure that no solution ever found contains incorrectly ordered pairs of elements. This means that also intermediate solutions contain some reliable information. Note that this cannot be realized with permutations only as search space.

We defer the detail to the following sections, but sketch the main concepts and results now. We shall not need all orders on the ground set $\mathcal{G}$ (which consists of the elements to be sorted) in our search space. Since we aim at a linear order on $\mathcal{G}$, we can restrict ourselves to orders that can be defined via assigning a predecessor to some elements of the set (meaning that the predecessor is 'smaller' than the element itself). In our case, this leads to directed forests in which each tree is directed towards a unique root. Since dealing with separate trees may be less convenient, we add an artificial element a_0 that is known to be smaller than all elements and arcs from the tree roots to this new element. This ensures that our search space can be represented by all directed trees on $\mathcal{G} \cup \{a_0\}$ such that the tree is directed towards its root a_0. As desired, this search space includes the empty order, represented by the tree having all elements of $\mathcal{G}$ as children of a_0 (cf. Figure 6.1), and all permutations, represented by trees that are simple directed paths ending in a_0 (cf. Figure 6.2).

This representation admits a natural elementary mutation: We choose two elements having the same predecessor (thus being sibling vertices in

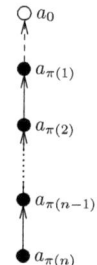

Figure 6.2: A permutation. If $a_{\pi(i)} \leq a_{\pi(i+1)}$ for all $1 \leq \pi(i) < n$ this is the optimal (sorted) solution.

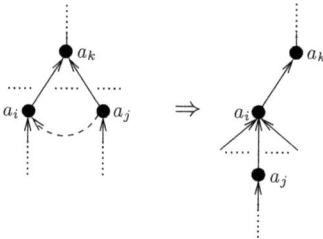

Figure 6.3: An elementary mutation.

the tree) and make the first one the new predecessor (i.e., the father) of the second one (cf. Figure 6.3). We present two probability distributions to choose the sibling vertices.

As fitness function, we use the number of correctly ordered element pairs which corresponds to the fitness function INV used in [STW04]. Since we aim at having no incorrectly ordered element pairs (an element having a predecessor that actually is larger than itself), we punish occurrences of such situations heavily.

As already discussed, this framework ensures that we shall only have correct ordering information about elements (meaning that if one element is the predecessor of another, than the first element is not larger than the other) in our population. We feel that this is a very desirable feature.

To analyze the algorithmic efficiency, we build a simple $(1 + 1)$ evolutionary algorithm from the components just discussed and analyze its optimization time. We will see that the optimization time and the number of element comparisons (the common measure when analyzing Sorting algorithms) have the same order of magnitude. We prove that our algorithm has with overwhelming probability and in expectation an optimization time of $O(n^2)$, where $n := |\mathcal{G}|$ is the number of elements to be sorted. Note that this is already faster than the current best evolutionary solution [STW04] having a proven expected optimization time of $O(n^2 \log n)$.

Our $O(n^2)$ bound is relatively robust with respect to the fitness function. It holds for any fitness function that punishes incorrectly ordered element pairs heavily, but does not punish finding new correctly ordered element pairs. Examples include the number of correctly ordered father-child pairs in the individual or the sum of distances between elements and their predecessors.

Next, we show that the optimization time is in fact a useful measurement of the efficiency of the algorithm, since each step of the algorithm can be computed efficiently (i.e., in constant or in logarithmic time, depending on which probability distribution we use to choose the sibling vertices for an elementary mutation).

On the experimental basis, our approach is even better than the $O(n^2)$ bound. We conduct several experiments that suggest that the expected optimization time is around $\Theta(n \log n)$. They clearly show that it is of smaller order than $n \log^2 n$. To have a comparison with the previous work, we also implemented their algorithms. The results indicate that the expected optimization time coincides with the proven upper bounds of $O(n^2 \log n)$.

6.2 An Evolutionary Algorithm for Sorting

Given a ground set $\mathcal{G} = \{a_1, \ldots, a_n\}$ and a total order $\leq$ on $\mathcal{G}$. The Sorting problem is the problem of finding an ordered sequence $(a_{\pi(1)}, \ldots a_{\pi(n)})$ such that $a_{\pi(i)} \leq a_{\pi(i+1)}$ for all $1 \leq \pi(i) < n$ where π is a permutation of $\{1, \ldots, n\}$. It is well known that any Sorting algorithm based on comparisons only needs in the worst case a runtime of $\Omega(n \log n)$.

We now introduce a representation of the individuals for the Sorting problem. From this representation we directly get a useful initial solution. A mutation operator for the individuals can also be derived from the representation. The definition of a fitness function completes the necessary components for a $(1 + 1)$ evolutionary algorithm.

6.2.1 Individuals

An ordered sequence, and thus an optimal solution to the Sorting problem, can be identified with a permutation of the elements of $\mathcal{G}$. However, the intermediate results of many Sorting algorithms cannot be represented this way. Thus, we consider a different search space that can represent a wider range of orders, namely directed forests where each component is directed towards its root and an arc (a_i, a_j) means $a_j \leq a_i$. This can nicely be achieved by assigning predecessors to some of the elements. Since it is more convenient to work with a single connected component instead of with different trees, we add an artificial element a_0 not belonging to $\mathcal{G}$ that is known to be smaller than all other elements. We then connect all tree roots by an arc to this new element a_0 (in other words, the predecessors of the tree roots is set to a_0). Thus, an individual I is represented by a vector of predecessors $I = (p(a_1), \ldots, p(a_n)) \in (\mathcal{G} \cup \{a_0\})^n$. Note that not all partial orders can be represented by such a tree, however the representation can help evolutionary algorithms to find linear orders more efficiently.

We aim at algorithms that successively find new ordering information (information that indicates the correct order of two elements) and add it to the current solution. Using our representation and an adequate fitness function (see below) it is easy to achieve that any intermediate solution contains for any two elements of $\mathcal{G}$ either the correct order (if $a_i \leq a_j$ than a_i is an ancestor of a_j in the tree) or no information. In the beginning we have no information, and thus instead of a random initial solution we use as initial solution I the empty order in which all elements a_i are unordered. This is represented by a tree in which the predecessor of every a_i for $1 \leq i \leq n$ is $p(a_i) = a_0$ (cf. Figure 6.1). This way, all elements of $\mathcal{G}$ are incomparable in I.

The final solution is a permutation of the elements of $\mathcal{G}$ which in the tree representation is a simple directed path ending in a_0 (cf. Figure 6.2). If for every arc $(a_{\pi(i+1)}, a_{\pi(i)})$ it holds that $a_{\pi(i)} \leq a_{\pi(i+1)}$, then this permutation is sorted.

6.2.2 Fitness Function

We propose one fitness function here and will show later that it can be replaced by several other fitness functions.

Since we search for a correct ordering of the n elements of $\mathcal{G}$, it is intuitive to give a positive reward (e.g., 1) for every correctly ordered pair of vertices

6.2 An Evolutionary Algorithm for Sorting

in the tree. Since we want to avoid incorrect orderings completely, we give a sufficiently high punishment (e.g., $-n^2$) for any wrong ordering of two elements. That way, we get the following fitness function $f\colon (\mathcal{G}\cup\{a_0\})^n \to \mathbb{Z}$ with

$$f(I) := \sum_{1 \leq i,j \leq n} f(a_i, a_j)$$

where

$$f(a_i, a_j) := \begin{cases} 1 & \text{if } a_i \leq a_j \text{ and } a_i = p^k(a_j) \text{ for some } 1 \leq k < n, \\ -n^2 & \text{if } a_i \leq a_j \text{ and } a_j = p^k(a_i) \text{ for some } 1 \leq k < n, \\ 0 & \text{otherwise} \end{cases}$$

and $p^k(a_j) = \overbrace{p(p(\ldots p(a_j)))}^{k} = a_i$ for some $1 \leq k < n$ if a_i is an ancestor of a_j in I.

This fitness function f has to be maximized. The value of f is 0 for the initial individual and $\frac{1}{2}n(n-1)$ for the optimal one. We will see in Section 6.3 that we can use a number of other (and possibly easier) fitness functions instead and in Section 6.4 how the fitness function can be computed efficiently.

6.2.3 Mutation Operator

A natural mutation operator for this representation will try to add information to the current solution by randomly assigning new predecessors to some vertices.

Hence, an elementary mutation is to assign a new predecessor to one of the vertices $a_j \in \mathcal{G}$. Since the ordering contained in an intermediate solution is correct and we do not wish to destroy correct ordering information, an elementary mutation picks two vertices a_i and a_j having the same father (two sibling vertices with $p(a_i) = p(a_j)$) and makes the first one the new father of the second $p(a_j) = a_i$ (cf. Figure 6.3).

We propose and use the following two probability distributions to choose the sibling vertices.

D1: Pick one of the fathers having at least two children uniformly at random and then pick two of its children uniformly at random.

D2: Choose a pair of sibling vertices uniformly at random from all pairs of sibling vertices.

Both probability distributions comply with the following property.

P1: The probability that one elementary mutation makes a_j the predecessor of a_i is the same as the probability that it makes a_i the predecessor of a_j.

As we have explained in Section 2.4, the mutation operator picks a number S at random according to a Poisson distribution $\text{Pois}(\zeta = 1)$ with parameter $\zeta = 1$. Hence, the probability that S is set to k is $\Pr[S = k] = \frac{1}{ek!}$. The mutation operator then repeats the elementary mutation described above $S + 1$ times on I.

6.2.4 The $(1 + 1)$-EA$_{\text{sort}}$

Given the described representation, mutation operator, and fitness function, the following $(1 + 1)$ evolutionary algorithm for Sorting (from now on called $(1 + 1)$-EA$_{sort}$) naturally arises from these components. The $(1 + 1)$-EA$_{sort}$ starts with the initial population consisting of the initial solution I described above. This solution I is modified by a mutation step to get a new solution I'. A selection step replaces I by I' if the fitness $f(I')$ of the new solution I' is not worse than the fitness of I. The mutation and selection steps are repeated until the optimal solution is found. Pseudocode for the $(1 + 1)$-EA$_{sort}$ for sorting n elements is given in Figure 6.4.

The selection step (together with the above described fitness function) assures that no solution containing incorrectly ordered vertex pairs is ever accepted.

The $(1 + 1)$-EA$_{sort}$ has several benefits over the $(1+1)$ evolutionary algorithm based on permutations proposed by Scharnow, Tinnefeld, and Wegener [STW04]. For one, using the tree representation our algorithm can also be used to find a wider range of partial orders. Another advantage is that even if the algorithm has not finished sorting the elements completely, the preliminary result is still useful, as for every pair of elements it either gives the correct order of the elements or no order. Last but not least, we will see that using this data structure we outperform the algorithms based on permutations.

6.3 Analysis of the Optimization Time

$(1+1)$-EA FOR SORTING
  Initialization:
1   $\mathcal{I} \leftarrow \{I\}$ where I is the empty order.
2   **repeat**
      Mutation:
3   Pick S according to $\Pr[S = k] = \frac{1}{e \cdot k!}$.
4       $I^0 \leftarrow I$
5   for $m = 1$ to $S+1$
6   do
7   Pick two sibling vertices a_i, a_j by **Dist**$_1$ or **Dist**$_2$.
8   Generate I^m from I^{m-1} by making a_i
9   the father of a_j.
      Selection:
10   if $f(I^{S+1}) \geq f(I)$
11   then $\mathcal{I} \leftarrow \{I^{S+1}\}$
12   **until** $\mathcal{I}$ contains the optimal solution

  Dist$_1$:
13   Choose an element having at least 2 children uniformly at random.
14   Choose 2 of the children uniformly at random.

  Dist$_2$:
15   Choose 2 sibling vertices uniformly at random from all pairs of sibling vertices.

Figure 6.4: Pseudocode for the $(1+1)$-EA$_{sort}$ for Sorting.

6.3 Analysis of the Optimization Time

In this section, we prove upper and lower bounds for the optimization time of the $(1+1)$-EA$_{sort}$. We show that an upper bound of $O(n^2)$ holds in expectation and also with overwhelming probability. A natural $\Omega(n \log n)$ lower bound is derived from classical theory. Finally, we show that both bounds in fact hold for any fitness function that rewards finding additional correct ordering information but forbids accepting wrong ordering information.

The following upper bound on the optimization time of the $(1+1)$-EA_{sort} holds, regardless of which of the two probability distributions D1 and D2 proposed in the previous section is used to choose the two elements for the mutation step.

Theorem 6.1. *As long as the elementary mutations comply with the property P1, the $(1+1)$-EA_{sort} needs in expectation and with overwhelming probability[1] at most $O(n^2)$ steps to find the optimal solution.*

Proof. Since P1 holds, for any two elements a_i and a_j the probability to choose a_i and a_j in an elementary mutation, and thus to make a_i the predecessor of a_j, is the same as the probability to choose a_j and a_i. Since one of the two elementary mutations increases the fitness by at least one and the other one is rejected, with a probability of $\frac{1}{2}$ an elementary mutation increases the fitness by at least one.

In the previous section we have seen that the fitness of the initial solution is 0 and the fitness of the optimal solution is $\frac{1}{2}n(n-1)$. If the $(1+1)$-EA_{sort} has not found the optimal solution yet, the fitness of the individual is smaller than $\frac{1}{2}n(n-1)$. Since every step that is accepted increases the fitness by at least 1, the fitness has then been increased by less then $\frac{1}{2}n(n-1)$ mutation steps. We show for any constant $\eta > e$ that if the $(1+1)$-EA_{sort} performs $\eta n(n-1)$ steps with overwhelming probability at least $\frac{1}{2}n(n-1)$ of the mutation steps would consist of a single elementary mutation and be accepted.

Let $t := \eta n(n-1)$ for some constant $\eta > e$ and let t' be the number of mutation steps the $(1+1)$-EA_{sort} needs to find the optimal solution. We count the number of accepted mutation steps that consist of a single elementary mutation. For that we define for $1 \leq i \leq t'$ the binary random variables X_i by $X_i = 1$ if the i-th mutation step of the $(1+1)$-EA_{sort} consists of a single elementary mutation and increases the fitness and $X_i = 0$ otherwise. The probability that the i-th mutation step consists of a single elementary mutation is $\frac{1}{e}$ by definition of the Poisson distribution, and the probability that the mutation step consisting of this elementary mutation is accepted is $\frac{1}{2}$. Thus independent of the steps before, $Pr[X_i = 1] = p := \frac{1}{2e}$ and $Pr[X_i = 0] = 1-p$. For $t' < i \leq t$, define the mutually independent random variables X_i by $Pr[X_i = 1] := p$ and $Pr[X_i = 0] := 1-p$. Then, all X_i are mutually independent.

The expected value of $X := \sum_{i=1}^{t} X_i$ is then $\mathbb{E}[X] = \frac{\eta}{2e}n(n-1)$.

---

[1] Recall that "with overwhelming probability" means that an event happens with probability at least $1 - 2^{-\Omega(n^\epsilon)}$ for a constant $\epsilon > 0$.

6.3 Analysis of the Optimization Time

Thus, if the $(1+1)$-EA$_{sort}$ has not found the optimal solution after $t = \eta n(n-1)$ steps, the fitness has been increased by less then $\frac{1}{2}n(n-1)$ mutation steps, and $X < \frac{1}{2}n(n-1)$. If we use $\alpha = \frac{e}{\eta} < 1$, we have that $\alpha \mathbb{E}[X] = \frac{1}{2}n(n-1)$. Thus, we can use the first inequality of Theorem 3.1 to bound the probability that the $(1+1)$-EA$_{sort}$ does not find the optimal solution in t steps by

$$\begin{aligned}
Pr\begin{bmatrix}\text{The } (1+1)\text{-EA}_{sort} \text{ does not find} \\ \text{the optimal solution in } t \text{ steps}\end{bmatrix} &\leq Pr[X < \tfrac{1}{2}n(n-1)] \\
&= Pr[X < \alpha\mathbb{E}[X]] \\
&< \exp\left(-\tfrac{1}{2}(1-\alpha)^2 \mathbb{E}[X]\right) \\
&= \exp\left(-\frac{(\eta-e)^2}{\eta^2} \frac{\eta}{4e} n(n-1)\right) \\
&= \exp\left(-\frac{(\eta-e)^2}{4e\eta} n(n-1)\right).
\end{aligned}$$

Thus, we have shown that the $(1+1)$-EA$_{sort}$ needs with overwhelming probability at most $O(n^2)$ mutation steps to find the optimal solution.

The expected upper bound follows from Lemma 3.3. □

Theorem 6.2. *In the worst case, the $(1+1)$-EA$_{sort}$ needs at least an expected number of $\Omega(n \log n)$ steps to find the optimal solution.*

Proof. A mutation step chooses a random number S drawn from a Poisson distribution Pois($\zeta = 1$) with parameter $\zeta = 1$ and performs $S+1$ elementary mutations, each changing the predecessor of one element. The selection step compares the fitness function values of the new and the old individual. However, it would suffice if the selection step compared only the elements for which the predecessor has changed with the new predecessor of these elements (cf. Section 6.4), thus performing one comparison per elementary mutation.

Since the expected value for a Poisson distribution is ζ, a mutation step performs in expectation two comparisons. Thus, if the $(1+1)$-EA$_{sort}$ applies t mutation steps to find the optimal solution, this corresponds in expectation to performing $2t$ comparisons. Since any randomized algorithm needs at least an expected number of $\Omega(n \log n)$ comparisons [CLRS03], the $(1+1)$-EA$_{sort}$ needs in expectation at least $\Omega(n \log n)$ mutation steps to find the optimal solution. □

This proof also reveals that the optimization time and the number of element comparisons, the common measure for analyzing Sorting algorithms, have the same order of magnitude.

Although the fitness function f given in Section 6.2 is the natural choice for a fitness function, it may not be the one that is easiest to analyze. However, we can easily choose a different fitness function, as long as the following two restrictions are followed.

R1: The fitness function awards additional correct ordering information with a non-negative reward.

R2: The fitness function prevents the acceptance of incorrect ordering information (e. g., by punishing the incorrect order of two elements heavily).

Lemma 6.1. *If the $(1+1)$-EA_{sort} uses instead of the original fitness function f a fitness function f' that complies with the restrictions R1 and R2, the behavior of the $(1+1)$-EA_{sort} is the same as if it uses f.*

Proof. An elementary mutation chooses two sibling vertices and makes one the father of the other. Differently put, it chooses two elements that are incomparable in the current solution and makes them comparable.

Consider a mutation step consisting of $S + 1 \geq 1$ elementary mutations. Assume one of the $S+1$ elementary mutations introduces incorrect ordering information in the current individual I, in other words it makes a_i the father of a_j and $a_i > a_j$. Now a_i and a_j are comparable in I. Since an elementary mutation always chooses sibling vertices, there is no way to make two comparable elements incomparable, and thus no other elementary mutation can destroy the effects of this elementary mutation. Thus, after the $S+1$ elementary mutations, a_i is still a predecessor of a_j and the newly created individual contains incorrect ordering information. Due to the restriction R2, both fitness function, f and f', reject the new individual.

Otherwise, only correct ordering information is introduced. Due to the restriction R1, the fitness function value of the new individual is not smaller than the fitness of the old individual, and thus the new individual is accepted using either of the two fitness functions.

Thus we have proven that if the original fitness function f accepts or rejects the new individual so does the alternative fitness function f'. □

Note that this lemma only guarantees that the acceptance behavior of the $(1+1)$-EA_{sort} is the same no matter whether it uses f or f'. Differently put,

$f(I') \geq f(I)$ if and only if $f'(I') \geq f'(I)$. However, the fitness values of f and f' can differ for the same individual. Hence Lemma 6.1 is not transferable to evolutionary algorithms working with more than one candidate solution in the population and one as offspring.

The following two fitness functions $f' \colon (\mathcal{G} \cup \{a_0\})^n \to \mathbb{Z}$ and $f'' \colon (\mathcal{G} \cup \{a_0\})^n \to \mathbb{R}$ are examples for fitness functions following the restrictions R1 and R2. The first example is

$$f'(I) := \sum_{1 \leq i,j \leq n} f'(a_i, a_j)$$

where

$$f'(a_i, a_j) := \begin{cases} 1 & \text{if } a_i \leq a_j \text{ and } a_i = p(a_j), \\ -n^2 & \text{if } a_i \leq a_j \text{ and } a_j = p(a_i), \\ 0 & \text{otherwise.} \end{cases}$$

This function f' counts the number of correctly ordered father-child pairs in the individual. The second function f'' can only be used if some distance metric is defined on the elements $\{a_0, \ldots, a_n\}$.

$$f''(I) := \sum_{1 \leq i,j \leq n} f''(a_i, a_j)$$

where

$$f''(a_i, a_j) := \begin{cases} |a_j - a_i| & \text{if } a_i \leq a_j \text{ and } a_i = p(a_j), \\ -n \cdot \max_{i,j \in [1..n]} \{|a_j - a_i|\} & \text{if } a_i \leq a_j \text{ and } a_j = p(a_i), \\ 0 & \text{otherwise.} \end{cases}$$

The fitness function f'' sums the distances between elements and their predecessors in an individual.

6.4 Implementation Details and Analysis of the Actual Runtime

In this section we show that each component of the $(1+1)$-EA$_{sort}$ can be computed highly efficiently. That implies that the bounds for the optimization time given in the previous section also hold up to a constant factor for the runtime if we choose the father of the two sibling vertices uniformly at

name	use, size, initialization
pred:	An array that contains a solution tree by storing for each element a_i its predecessor $p(a_i)$. $n, [a_0, \ldots, a_0]$
father:	An array that contains in arbitrary order the elements having at least 2 children. $n, [0, -1, \ldots, -1]$
#child:	An array that contains for every element the number of its children. $n+1, [n, 0, \ldots, 0]$
child$_i$:	$n+1$ arrays that contain for every element a_i for $i \in [0..n]$ its children in arbitrary order. $(n+1) \cdot n, [a_1, \ldots, a_n]$ for $i = 0$ and $[-1, \ldots, -1]$ for $i \in [1..n]$
#desc:	An array that contains for every element the number of its descendants. $n+1, [n, 0, \ldots, 0]$

Figure 6.5: The datastructures used by the $(1+1)$-EA$_{sort}$ if the elementary mutations use probability distribution D1.

random (according to probability distribution D1). If we choose the two sibling vertices uniformly at random from all pairs of sibling vertices (according to probability distribution D2), the runtime increases by a factor of $O(\log n)$.

Assume first that the elementary mutations of the $(1+1)$-EA$_{sort}$ use the probability distribution D1. The first step of the algorithm is the initialization. For that, it creates and initializes the datastructures listed in Figure 6.5. Additionally, it uses counters for **father** and **child$_i$** to keep track of how many elements the arrays currently contain. This is done once and since the accumulated size of initial entries into the datastructures is $O(n)$ it needs $O(n)$ time.

To avoid the creation of a new array **pred** in every step of the algorithm, we keep two copies of **pred**, one holding the current individual I and the other one holding the newly created individual I'.

Then, the $(1+1)$-EA$_{sort}$ performs a number of mutation and selection steps. Each mutation step consists in expectation of two elementary mutations (cf. the previous section). For each elementary mutation, it first chooses randomly one of the fathers having at least two children (using the array **father** and the corresponding counter). Let this father be element a_k. Using the array **#child**, the $(1+1)$-EA$_{sort}$ can determine how many pairs of

6.4 Implementation Details and Analysis of the Actual Runtime

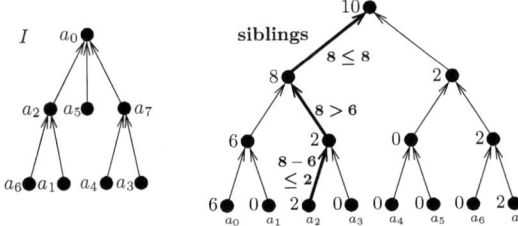

Figure 6.6: The figure shows an individual I (on the left) and its corresponding datastructure **siblings** (on the right). Most of the elements have no children (no sibling pairs), a_0 has 3 children (6 sibling pairs), and a_2 and a_7 have 2 children (2 sibling pairs) each. Thus, in total there are 10 sibling pairs. If an elementary mutation chooses the 8-th sibling pair, it descends the tree **siblings** as indicated in bold. It finds that this is the second of the sibling pairs being children of a_2.

sibling vertices are a_k's children. After it has chosen a random one of these pairs, the corresponding pair (a_i, a_j) can be found with help of the array **child$_k$**. Then, the algorithm sets $p(a_j) = a_i$ in the copy of **pred** corresponding to I'. All of the above and the updates of the arrays **father**, **#child**, **child$_k$**, and **#desc** can be done in constant time (for **father** and **child$_l$** the constant time can be achieved because the elements are unordered and the counters store how many elements are in the arrays). Hence, in expectation, a mutation step can be performed in constant time.

The fitness function can be computed in linear time by doing a depth-first traversal of the tree during which the tree-levels of the elements are summed up. If an incorrect ordering of some vertex and its father is found, a negative number is returned. However, by remembering which elementary mutations were performed, instead of recomputing the fitness function in every step, we can update the fitness function in constant time using the information stored in **#desc**.

Depending on the fitness values of the individuals I and I', the selection step either ignores I' or it replaces I by I'. In the first case, we simply set the copy of **pred** for I' to the values of the copy of **pred** for I to be ready for the next mutation step. Since we remembered which elementary mutations were performed for the efficient computation of the fitness function, this can be done in expected constant time. Otherwise, the selection step replaces I by I' by updating the datastructures **father**, **#child**, **child$_l$**, **#desc**, and the copy of **pred** for I. Again, these updates can be done in expected constant time.

Now assume the probability distribution D2 is used for the elementary mutations. We use the datastructure **siblings** instead of **father**. **siblings** is a complete binary tree of size $2^{\lceil \log_2(n+1) \rceil}$. The i-th leaf is associated with element a_i for $i \in [0..n]$ and the label of this leaf holds the number of pairs of sibling vertices that are sons of element a_i in the current solution I. The internal vertices hold the sum of the values of its two sons. Thus, a label of an internal vertex u holds the number of pairs of sibling vertices that are children of one of the elements associated with a leaf in u's subtree (cf. Figure 6.6). Consequently, the label of the root of the tree **siblings** holds the total number of pairs of sibling vertices in I.

Since only the computation of the sibling pair on which an elementary mutation is performed has changed, we only have to consider this computation and the initialization and the updates on the datastructure **siblings**. The tree **siblings** is initialized by setting the label of the leaf associated with a_0 to $n(n-1)$ which is the number of possible sorted pairs of n elements. The labels of the ancestors of this leaf are also set to $n(n-1)$, all other labels are set to 0. This has to be done once and can be achieved in linear time. D2 chooses one of all pairs of sibling vertices at random. For that it chooses a number between 1 and the number of sibling pairs in the individual (given by the label of the root of **siblings**). To find the corresponding pair, it descends the tree **siblings** according to the labels (cf. Figure 6.6) until it has found the correct father a_k. Using **child$_k$**, the $(1+1)$-EA$_{sort}$ can compute the correct sibling pair. Descending the tree **siblings** and its subsequent update can be done in O($\log n$) time. Thus, a mutation step consisting of an expected number of two elementary mutations (which use the D2) can be performed in expected logarithmic time.

Hence, we have shown that a step of the $(1+1)$-EA$_{sort}$ can be performed in expected constant time if the elementary mutations use the probability distribution D1. If the elementary mutations use instead probability distribution D2, a step of the algorithm can be performed in expected logarithmic time.

6.5 Experimental Results

We implemented the $(1+1)$-EA$_{sort}$ described in the previous sections as well as the evolutionary algorithm for Sorting presented in [STW04] (in this section referred to as $(1+1)$-EA$_\text{p}$ since it uses permutations as representation). This section contains the results of the experiments we carried out using both implementations.

6.5 Experimental Results

The bounds we proved in Section 6.3 are not tight. However, they already show that the expected optimization time of our $(1+1)$-EA$_{sort}$ using a tree representation is at least as good as the expected optimization time of the $(1+1)$-EA$_p$ given in [STW04]. Our upper bound is $O(n^2)$, whereas the general lower bound of the $(1+1)$-EA$_p$ is $\Omega(n^2)$ and $\Omega(n^2 \log n)$ for several of the combinations of fitness function and mutation operator (in particular those combinations that seem most adequate). We will see that the real expected optimization time of our algorithm seems to be much closer to the lower bound of $\Omega(n \log n)$ than to the upper bound. The expected optimization time of the $(1+1)$-EA$_p$ appears to be close to $\Theta(n^2 \log n)$ as the proven bounds indicate.

As described in Section 6.2, an elementary mutation of our evolutionary algorithm chooses two elements having the same father and makes the first one the father of the second. The two probability distributions we proposed to choose these elements are choosing the father of the two elements uniformly at random (D1) or choosing a sibling pair uniformly at random from all pairs of sibling vertices (D2). We implemented both variants of the algorithm and conducted test runs. For each variant we let the algorithm solve 100 instances for the input sizes $100, 300, \ldots, 5900$. The results (cf. Figure 6.7) clearly indicate that the average optimization time of our algorithm is only slightly higher than linear in the number of elements to be sorted. The smaller graphs in the figure shows the same optimization times divided by $n \log n$ in the first graph and divided by $n \log^2 n$ in the second graph. These graphs show that the real expected optimization time seems to lie between $\Omega(n \log n)$ and $O(n \log^2 n)$.

The $(1+1)$-EA$_p$ to which we compare our algorithm comes in many flavors. For one, 5 different fitness functions are used (cf. also Section 6.1.1), namely HAM, EXC, INV, LAS, and RUN. Additionally, two different mutation operators, JUMP and EXCHANGE, were proposed.

First we wanted to find out for each fitness function whether it is best to use both mutation operators or only one of them, and if so which one. To this aim, we let the $(1+1)$-EA$_p$ solve for each fitness function except RUN 10 random instances for the input sizes $10, 20, \ldots 100$ either using only JUMP or only EXCHANGE or using both with probability $\frac{1}{2}$ each. Since RUN has a very high optimization time, we chose to let this variant of the $(1+1)$-EA$_p$ solve 10 random instances for the input sizes $2, 4, \ldots, 20$. The results (cf. Figure 6.8) indicate that for the following experiments we should choose the mutation operator EXCHANGE for the fitness functions HAM and EXC and for the fitness functions LAS and RUN only JUMP. The fitness function INV is the only fitness function where all three variants are comparable, however

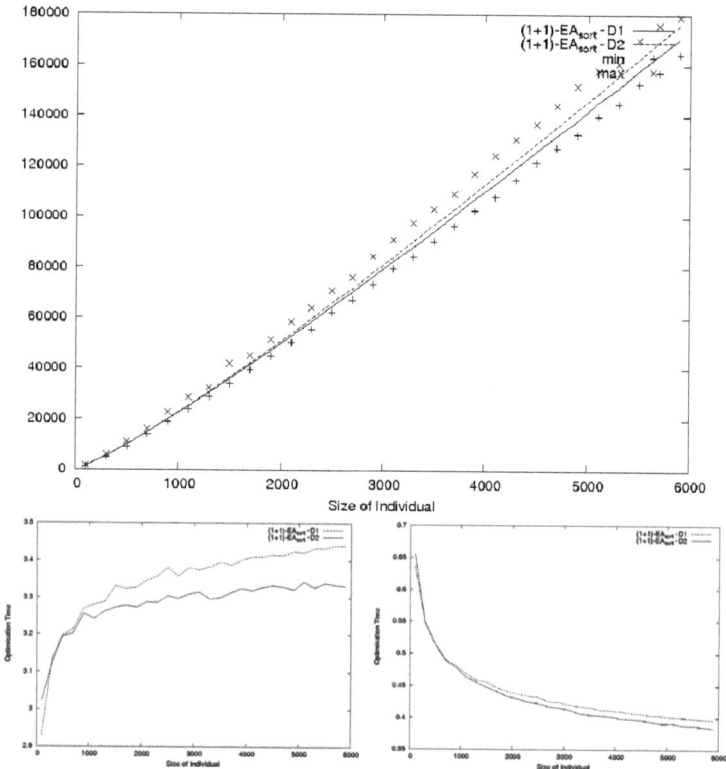

Figure 6.7: The main graph shows the optimization times for the $(1+1)$-EA$_{sort}$ using the probability distributions D1 and D2 for the elementary mutations, averaged over 100 runs each. Additionally the minimum and maximum optimization times over all 200 runs are given. The smaller graphs show the average optimization times divided by $n \log n$ (left graph) and $n \log^2 n$ (right graph). These plots indicate that the actual expected optimization time is around $\Theta(n \log n)$.

using only JUMP seems to be slightly superior. Note that the tight bounds in [STW04] were given for HAM and EXC using only the EXCHANGE operator and for INV using either operator, which coincides with the best mutation operators we determined for these fitness functions.

6.5 Experimental Results

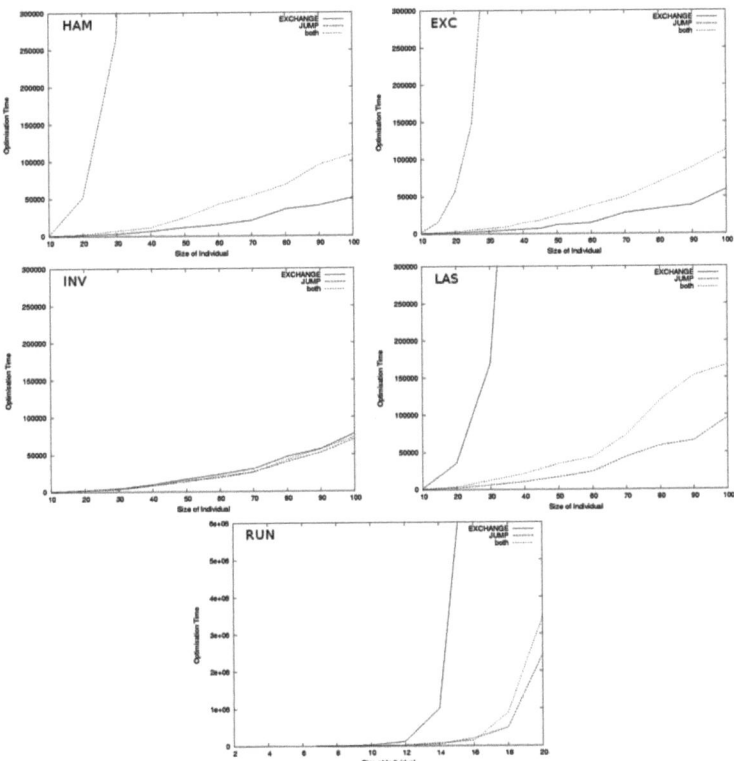

Figure 6.8: The first graph shows the average optimization times of the $(1+1)$-EA_p using the fitness function HAM and either only the JUMP or only the EXCHANGE operator or both. The next graphs show the same for the $(1+1)$-EA_p using the fitness functions EXC, INV, LAS, and RUN.

Having found the adequate mutation operators for each fitness function, we chose to first have 10 runs for each fitness function on input sizes $\{1, 2, \ldots, 19\}$. Since RUN seems to have an expected exponential optimization time, we then let the $(1+1)$-EA_p for the other fitness functions make 10 runs of the input sizes between 10 and 300 with a step size of 10. As can easily be seen (cf. Figure 6.9), RUN has a very high and probably exponential expected runtime, as suggested in [STW04]. The others seem to have an

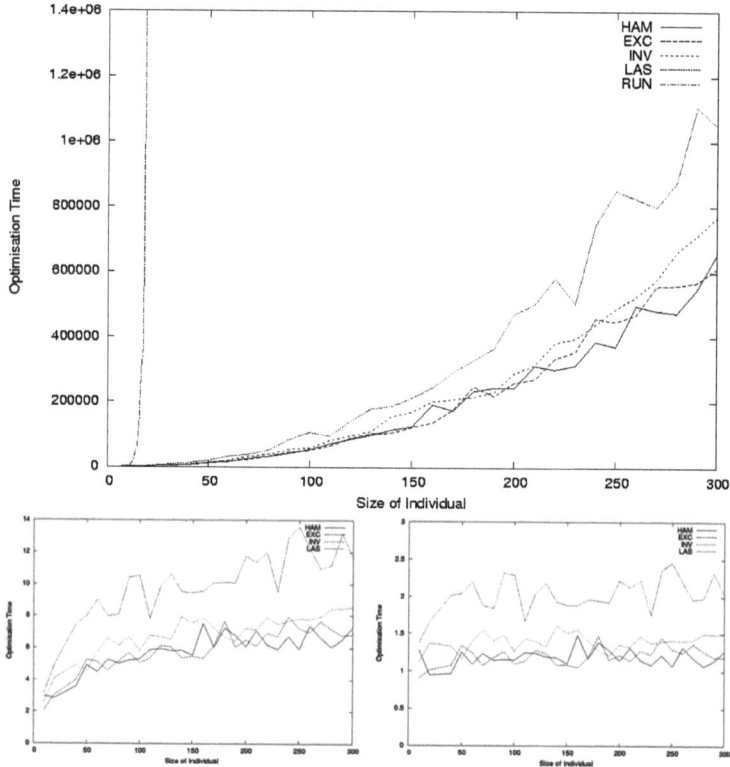

Figure 6.9: The main graph shows the average optimization times of the $(1+1)$-EA_p for all fitness functions using the best mutation operator for the respective fitness function. The smaller graphs show the same optimization times (excepting the times for RUN) divided by n^2 (left graph) and by $n^2 \log n$ (right graph).

expected optimization time of $O(n^2 \log n)$.

Finally, we directly compare our $(1+1)$-EA_{sort} and the $(1+1)$-EA_p. For this, we let the $(1+1)$-EA_{sort} (with both probability distributions used for the elementary mutations) and the $(1+1)$-EA_p (with the fitness functions HAM, EXC, INV, and LAS and the according mutation operators) solve 10 random instances for the input sizes $10, 20, \ldots, 200$ (cf. Figure 6.10). Note

6.5 Experimental Results

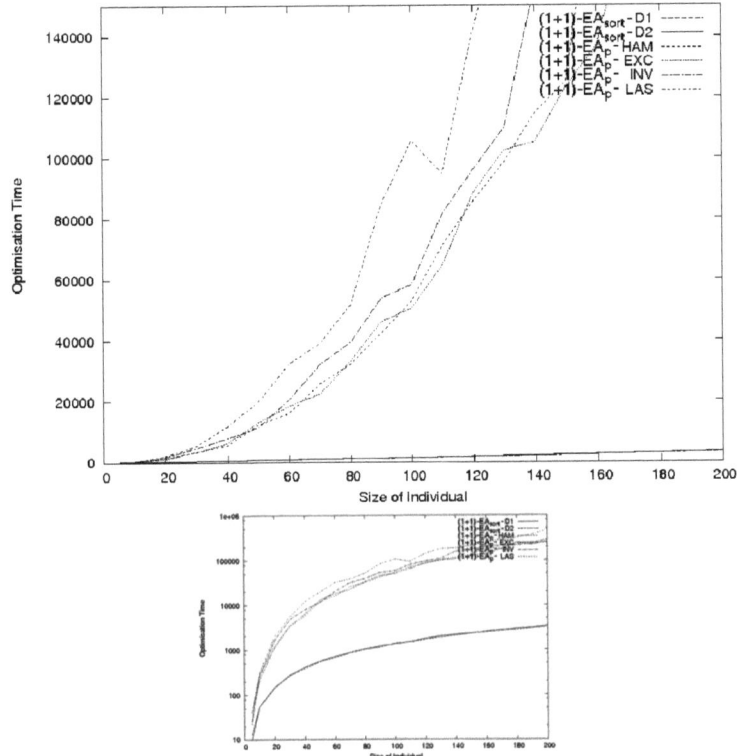

Figure 6.10: The main graph shows the average optimization times of the $(1+1)$-EA$_{sort}$ versus the average optimization times of the $(1+1)$-EA$_p$. The smaller one shows the same optimization times scaling the y-axis logarithmically.

that even using the logarithmic scale it is barely possible to distinguish the two variants of the $(1+1)$-EA$_{sort}$. The comparison clearly shows that our algorithm outclasses the $(1+1)$-EA$_p$. Allowing, e.g., an optimization time of 50000, all variants of th $(1+1)$-EA$_p$ can only handle instances with $n \leq 100$ whereas our algorithm works up to $n \leq 2000$ (cf. Figures 6.7 and 6.8).

6.6 Summary

In this chapter, we suggested a novel approach to problems where the final solution is a permutation (linear order). When applied to the problem of sorting n comparable elements, it has the structural advantage of disallowing incorrect ordering information. A simple $(1+1)$ evolutionary algorithm built upon this framework was faster than previous such algorithms both theoretically (proven asymptotic bounds) and experimentally. A problem left open in this chapter is giving a mathematical proof that the evolutionary algorithm proposed indeed has an expected optimization time of $\Theta(n \log n)$ as observed in the experiments. We are optimistic that our approach via orders instead of permutations will also be useful for other (non-linear) order problems.

Chapter 7
Summary

This work contains runtime analyses for evolutionary algorithms for three prominent problems from computer science, i.e., the Single Source Shortest Path problem, the All-Pairs Shortest Path problem, and the Sorting problem. For the analyses, we developed and used several probability theoretical tools that will probably be helpful for future analyses of such algorithms.

For the Single Source Shortest Path problem, we analyzed an already known and analyzed $(1+1)$ evolutionary algorithm. Devising a new method for the analysis, we gave a tight runtime analysis for this algorithm. Following this, we designed a natural evolutionary algorithm for the All-Pairs Shortest Path problem. We proved that adding a natural crossover operator to this algorithm provably reduces the runtime. Experimental studies reveal that this is already the case for small instances. This is the first time that the usefulness of crossover could be shown for a non-artificial problem. Last, we designed a new representation based on trees for Sorting / Ordering problems. We showed that the $(1+1)$ evolutionary algorithm for Sorting that naturally arises from this representation is provably faster than previous evolutionary algorithms for this problem. Experiments demonstrate that this improvement is evident even on small instances. Our new representation additionally has the advantage that it distinguishes between correct and unknown information. This way, even before the algorithm has finished, one can already extract a partial, correct solution.

Appendix A
Further Contributions

The main contributions of my time as PhD student have been thoroughly described in this theses. However, during my studies I have additionally worked on the following two topics.

- **Virtual Private Network Design:** Virtual private network design deals with the reservation of capacities in a weighted graph such that the terminals in this network can communicate with one another. Each terminal is equipped with an upper bound on the amount of traffic that the terminal can send or receive. The task is to install capacities at minimum cost and to compute paths for each unordered terminal pair such that each valid traffic matrix can be routed along those paths.

  We considered a variant of the virtual private network design problem which generalizes the previously studied symmetric and asymmetric case. In our model the terminal set is partitioned into a number of groups, where terminals of each group do not communicate with each other. For this problem we designed an approximation algorithm which has an approximation guarantee of 4.74.

  "Provisioning a Virtual Private Network Under the Presence of Non-Communicating Groups" by Friedrich Eisenbrand and E. H. [EH06]

- **Fitness-Proportional Selection for Evolutionary Algorithms:** Rigorous runtime analyses of evolutionary algorithms (EAs) mainly investigate algorithms that use elitist selection strategies. Two algorithms commonly studied are Randomized Local Search (RLS) and the $(1+1)$-EA and it is well known that both optimize any linear pseudo-Boolean function on n bits within an expected number of $O(n \log n)$

fitness evaluations. In this paper, we analyze variants of these algorithms that use fitness proportional selection.

A well-known method in analyzing the local changes in the solutions of RLS is a reduction to the gambler's ruin problem. We extend this method in order to analyze the global changes imposed by the $(1+1)$-EA. By applying this new technique we show that with high probability using fitness proportional selection leads to an exponential optimization time for any linear pseudo-Boolean function with non-zero weights. Even worse, all solutions of the algorithms during an exponential number of fitness evaluations differ with high probability in linearly many bits from the optimal solution.

Our theoretical studies are complemented by experimental investigations which confirm the asymptotic results on realistic input sizes.

"Rigorous Analyses of Fitness-Proportional Selection for Optimizing Linear Functions" by E. H., D. Johannsen, C. Klein, and F. Neumann [HJKN08]

Bibliography

[AR02] C. W. Ahn and R. S. Ramakrishna. A genetic algorithm for shortest path routing problem and the sizing of populations. *IEEE Transactions on Evolutionary Computation*, 6:566–579, 2002.

[AS00] N. Alon and J. Spencer. *The Probabilistic Method*. John Wiley, 2nd edition, 2000.

[BT96] T. Blickle and L. Thiele. A comparison of selection schemes used in evolutionary algorithms. *Applied Soft Computing*, 4:361–394, 1996.

[CLRS03] T. H. Cormen, C. E. Leiserson, R. L. Rivest, and C. Stein. *Introduction to Algorithms*. McGraw-Hill Science / Engineering / Math, 2nd edition, 2003.

[DH08] B. Doerr and E. Happ. Directed trees: A powerful representation for sorting and ordering problems. In *Proceedings of the 2008 IEEE Congress on Evolutionary Computation (CEC)*, pages 3606–3613. IEEE Press, 2008.

[DHK07] B. Doerr, E. Happ, and C. Klein. A tight bound for the (1+1)-EA on the single source shortest path problem. In *Proceedings of the 2007 IEEE Congress on Evolutionary Computation (CEC)*, pages 1890–1895. IEEE Press, 2007.

[DHK08] B. Doerr, E. Happ, and C. Klein. Crossover can provably be useful in evolutionary computation. In *Proceedings of the 2008 Conference on Genetic and Evolutionary Computation (GECCO)*, pages 539–546. ACM Press, 2008.

[DHN06] B. Doerr, N. Hebbinghaus, and F. Neumann. Speeding up evolutionary algorithms through restricted mutation operators. In

Proceedings of the 9th International Conference on Parallel Problem Solving From Nature (PPSN), Lecture Notes in Computer Science, pages 978–987. Springer, 2006.

[Dij59] E. W. Dijkstra. A note on two problems in connexion with graphs. In *Numerische Mathematik*, volume 1, pages 269–271. Springer, 1959.

[DJ07] B. Doerr and D. Johannsen. Adjacency list matchings — an ideal genotype for cycle covers. In *Proceedings of the 2007 Conference on Genetic and Evolutionary Computation (GECCO)*, pages 1203–1210. ACM Press, 2007.

[DJW02] S. Droste, T. Jansen, and I. Wegener. On the analysis of the (1+1) evolutionary algorithm. *Theoretical Computer Science*, 276:51–81, 2002.

[DKS07] B. Doerr, C. Klein, and T. Storch. Faster evolutionary algorithms by superior graph representation. In *Proceedings of the First IEEE Symposium on Foundations of Computational Intelligence (FOCI)*, Lecture Notes in Computer Science, pages 245–250. IEEE Press, 2007.

[EH06] F. Eisenbrand and E. Happ. Provisioning a virtual private network under the presence of non-communicating groups. In *Proceedings of the 6th International Conference on Algorithms and Complexity (CIAC)*, Lecture Notes in Computer Science, pages 105–114. Springer, 2006.

[Flo62] R. W. Floyd. Algorithm 97: Shortest path. *Communications of the ACM*, 5:345, 1962.

[For93] S. Forrest. Genetic algorithms: Principles of natural selection applied to computation. *Science*, 261:872–878, 1993.

[FW04] S. Fischer and I. Wegener. The Ising model on the ring: Mutation versus recombination. In *Proceedings of the 2004 Conference on Genetic and evolutionary computation (GECCO)*, Lecture Notes in Computer Science, pages 1113–1124. Springer, 2004.

[GD91] D. E. Goldberg and K. Deb. A comparative analysis of selection schemes used in genetic algorithms. In *Foundations of Genetic Algorithms (FOGA)*, pages 69–93. Morgan Kaufmann, 1991.

BIBLIOGRAPHY

[GJ79] M. R. Garey and D. S. Johnson. *Computers and Intractability; A Guide to the Theory of NP-Completeness*. W. H. Freeman & Co., 1979.

[GW03] O. Giel and I. Wegener. Evolutionary algorithms and the maximum matching problem. In *Proceedings of the 20th Annual Symposium on Theoretical Aspects of Computer Science (STACS)*, Lecture Notes in Computer Science, pages 415–426. Springer, 2003.

[HJKN08] E. Happ, D. Johannsen, C. Klein, and F. Neumann. Rigorous analyses of fitness-proportional selection for optimizing linear functions. In *Proceedings of the 2008 Conference on Genetic and Evolutionary Computation (GECCO)*, pages 953–960. ACM Press, 2008.

[Hol75] J. H. Holland. *Adaption in Natural and Artificial Systems*. University of Michigan Press, 1975.

[HY04] J. He and X. Yao. A study of drift analysis for estimating computation time of evolutionary algorithms. *Natural Computing*, 3:21–35, 2004.

[IHK99] J. Inagaki, M. Haseyama, and H. Kitajima. A genetic algorithm for determining multiple routes and its applications. In *Proceedings of the 1999 IEEE International Symposium on Circuits and Systems (ISCAS)*, pages 137–140. IEEE Press, 1999.

[Joh77] D. B. Johnson. Efficient algorithms for shortest paths in sparse networks. *Journal of the ACM*, 24:1–13, 1977.

[JW99] T. Jansen and I. Wegener. On the analysis of evolutionary algorithms – a proof that crossover really can help. In *Proceedings of the 7th Annual European Symposium on Algorithms (ESA)*, Lecture Notes in Computer Science, pages 184–193. Springer, 1999.

[JW02] T. Jansen and I. Wegener. The analysis of evolutionary algorithms – a proof that crossover really can help. *Algorithmica*, 34:47–66, 2002.

[JW05] T. Jansen and I. Wegener. Real royal road functions – where crossover provably is essential. *Discrete Applied Mathematics*, 149:111–125, 2005.

BIBLIOGRAPHY

[KGJV83]   S. Kirkpatrick, D. Gelatt Jr., and M. P. Vecchi. Optimization by simulated annealing. *Science*, 220:671–680, 1983.

[LZHH02]   S. Liang, A. N. Zincir-Heywood, and M. I. Heywood. Intelligent packets for dynamic network routing using distributed genetic algorithm. In *Proceedings of the 2002 Conference on Genetic and Evolutionary Computation (GECCO)*, pages 88–96. Morgan Kaufmann, 2002.

[LZHH06]   S. Liang, A. N. Zincir-Heywood, and M. I. Heywood. Adding more intelligence to the network routing problem: AntNet and Ga-agents. *Applied Soft Computing*, 6:244–257, 2006.

[MHF93]   M. Mitchell, J. H. Holland, and S. Forrest. When will a genetic algorithm outperform hill climbing? In *Proceeding of the 7th Neural Information Processing Systems Conference (NIPS)*, Advances in Neural Information Processing Systems, pages 51–58. Morgan Kaufmann, 1993.

[MN99]   K. Mehlhorn and S. Näher. *LEDA: A Platform for Combinatorial and Geometric Computing*. Cambridge University Press, 1999.

[MR95]   R. Motwani and P. Raghavan. *Randomized Algorithms*. Cambridge University Press, 1995.

[MRR$^+$53]   N. Metropolis, A. W. Rosenbluth, M. N. Rosenbluth, A. H. Teller, and E. Teller. Equations of state calculations by fast computing machines. *Journal of Chemical Physics*, 21:1087–1091, 1953.

[Neu04]   F. Neumann. Expected runtimes of evolutionary algorithms for the Eulerian cycle problem. In *Proceedings of the 2004 IEEE Congress on Evolutionary Computation (CEC)*, pages 904–910. IEEE Press, 2004.

[NW04]   F. Neumann and I. Wegener. Randomized local search, evolutionary algorithms, and the minimum spanning tree problem. In *Proceedings of the 2004 conference on Genetic and evolutionary computation (GECCO)*, Lecture Notes in Computer Science, pages 713–724. Springer, 2004.

[NW05]   F. Neumann and I. Wegener. Minimum spanning trees made easier via multi-objective optimization. In *Proceedings of the 2005 Conference on Genetic and Evolutionary Computation (GECCO)*, pages 763–769. ACM Press, 2005.

BIBLIOGRAPHY

[PM95]   O. Petersson and A. Moffat. A framework for adaptive sorting. *Discrete Applied Mathmatics*, 59:153–179, 1995.

[RRS95] Y. Rabani, Y. Rabinovich, and A. Sinclair. A computational view of population genetics. In *Proceedings of the 27-th annual ACM symposium on Theory of computing (STOC)*, pages 83–92. ACM Press, 1995.

[RW91]   Y. Rabinovich and A. Wigderson. An analysis of a simple genetic algorithm. In *Proceedings of the 4th International Conference on Genetic Algorithms (ICGA)*, pages 215–221, 1991.

[RW99]   Y. Rabinovich and A. Wigderson. Techniques for bounding the convergence rate of genetic algorithms. *"Random Structures & Algorithms"*, 14:111–138, 1999.

[STW04] J. Scharnow, K. Tinnefeld, and I. Wegener. The analysis of evolutionary algorithms on sorting and shortest paths problems. *Journal of Mathematical Modelling and Algorithms*, 3:349–366, 2004.

[Sud05] D. Sudholt. Crossover is provably essential for the Ising model on trees. In *Proceedings of the 2005 conference on Genetic and evolutionary computation (GECCO)*, pages 1161–1167. ACM Press, 2005.

[SW04]   T. Storch and I. Wegener. Real royal road functions for constant population size. *Theoretical Computer Science*, 320:123–134, 2004.

[War62] S. Warshall. A theorem on boolean matrices. *Journal of the ACM*, 9:11–12, 1962.

[Weg01] I. Wegener. Theoretical aspects of evolutionary algorithms. In *Proceedings of the 28th International Colloquium on Automata, Languages and Programming (ICALP)*, Lecture Notes in Computer Science, pages 64–78. Springer, 2001.

[Weg03] I. Wegener. Towards a theory of randomized search heuristics. In *Proceedings of the 28th International Symposium on Mathematical Foundations of Computer Science (MFCS)*, Lecture Notes in Computer Science, pages 125–141. Springer, 2003.

[Wit05] C. Witt. Worst-case and average-case approximations by simple randomized search heuristics. In *Proceedings of the 22nd*

Annual Symposium on Theoretical Aspects of Computer Science (STACS), Lecture Notes in Computer Science, pages 44–56. Springer, 2005.

[WM97] D. H. Wolpert and W. G. Macready. No free lunch theorems for optimization. *IEEE Transactions on Evolutionary Computation*, 1:67–82, 1997.

[WW03] I. Wegener and C. Witt. On the optimization of monotone polynomials by simple randomized search heuristics. *Combinatorics, Probability and Computing*, 14:225–247, 2003.

[WW05] I. Wegener and C. Witt. On the optimization of monotone polynomials by simple randomized search heuristics. *Combinatorics, Probability and Computing*, 14:225–247, 2005.

yes
I want morebooks!

Buy your books fast and straightforward online - at one of the world's fastest growing online book stores! Environmentally sound due to Print-on-Demand technologies.

Buy your books online at
www.get-morebooks.com

Kaufen Sie Ihre Bücher schnell und unkompliziert online – auf einer der am schnellsten wachsenden Buchhandelsplattformen weltweit! Dank Print-On-Demand umwelt- und ressourcenschonend produziert.

Bücher schneller online kaufen
www.morebooks.de

OmniScriptum Marketing DEU GmbH
Heinrich-Böcking-Str. 6-8
D - 66121 Saarbrücken
Telefax: +49 681 93 81 567-9

info@omniscriptum.com
www.omniscriptum.com

Printed by Books on Demand GmbH, Norderstedt / Germany